The biochemistry of viruses

Cambridge Texts in Chemistry and Biochemistry

GENERAL EDITORS

D. T. Elmore
Professor of Biochemistry
The Queens's University of Belfast

J. Lewis
Professor of Inorganic Chemistry
University of Cambridge

K. Schofield
Professor of Organic Chemistry
University of Exeter

J. M. Thomas
Professor of Physical Chemistry
University of Cambridge

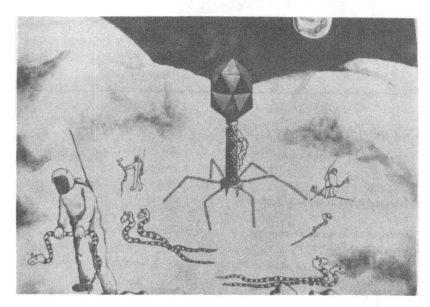

The invaders of the genosphere.

The biochemistry of viruses

S. J. MARTIN

Department of Biochemistry
Queen's University of Belfast, N. Ireland

CAMBRIDGE UNIVERSITY PRESS

CAMBRIDGE
LONDON NEW YORK NEW ROCHELLE
MELBOURNE SYDNEY

CAMBRIDGE UNIVERSITY PRESS
Cambridge, New York, Melbourne, Madrid, Cape Town, Singapore,
São Paulo, Delhi, Dubai, Tokyo, Mexico City

Cambridge University Press
The Edinburgh Building, Cambridge CB2 8RU, UK

Published in the United States of America by Cambridge University Press, New York

www.cambridge.org
Information on this title: www.cambridge.org/9780521292290

First published 1978
Reprinted 1979
Re-issued 2010

A catalogue record for this publication is available from the British Library

Library of Congress Cataloguing in Publication Data
Martin, Samuel John, 1936-
The biochemistry of viruses.
(Cambridge texts in chemistry and biochemistry)
Bibliography: p. 000
Includes index.
1. Viruses. 2. Microbiological chemistry.
I. Title. II. Series. [DNLM: 1. Viruses. 2. Virus diseases. QW160 M383b]
QR467.M37 576'.64 77-8231

ISBN 978-0-521-21678-4 Hardback
ISBN 978-0-521-29229-0 Paperback

Contents

v

Preface

This book is principally designed to provide a rapid overall picture of virology at the molecular level and should be of interest to students of biochemistry, biology, microbiology and the physical sciences who have had no formal training in virology. It should also be of interest to medical and veterinary students as it outlines the basic principles involved in virus diseases of man, animals and plants.

However, this short book is not intended as a comprehensive text in virology as little attention had been given to genetic, epidemiological or pathological problems. Rather, my aim has been to describe in a simple and comparative manner the variety of structures found in viruses and to compare the different strategies that viruses adopt when they infect cells and animals.

Although I have considered viruses as molecules, they do possess many of the basic features of organisms particularly in their ability to replicate. I have therefore threaded through the text a slightly philosophical theme and have attempted to compare some of the activities of viruses to those of Man. Man is the invader of the biosphere whereas viruses are the invaders of the genosphere.

Finally it is a pleasure to thank the many colleagues who have made this book possible. In particular, I thank Dr F. Brown (The Animal Virus Research Institute, Pirbright) for critically reading the manuscript. The electron micrographs were kindly provided by Dr E. Dermott (Department of Microbiology and Immunology, Queen's University, Belfast), Dr B. Adair (Veterinary Research Laboratories, Stormont, Belfast), Dr P. Cooper (Department of Plant Pathology, Queen's University, Belfast) and Mr C. Smale (Animal Virus Research Institute, Pirbright). I also thank Mrs Sandra Turley for painting the cartoon 'Invaders of the Genosphere'.

I am also grateful for many useful discussions with present and past members of my research group, in particular for discussions with Dr E. Hoey and Dr B. Rima concerning various aspects of the script; not least I am grateful to many of the Honours students in Biochemistry at Queen's who have over the years rewarded me by showing an interest in and enthusiasm for virology.

Belfast, November 1976   S.J.M.

1 A brief history of virology

1.1. The discovery of viruses

During the last few decades, the term virus has become a household word, largely because so many common human and animal diseases have been shown to be caused by viruses. Virus diseases are diseases of populations, since it is within closely knit communities of humans, animals or plants that viruses are able to infect, spread and flourish. Hence, as we now face a future of increasing population, intensive husbandry of farm animals and the apparently inevitable spread of the monoculture system of crop production, it is becoming increasingly important that we understand and thereby hopefully control the devastation that virus epidemics can so obviously cause.

During the last century it was established that many diseases are the result of infection by bacteria or protozoan cells. Bacteria were the smallest cells then known and could be seen in an ordinary light microscope. Also, the early bacteriologists had devised methods of isolating and growing bacterial pathogens on simple nutrient solutions. Other diseases were known, however, that could not be explained by bacterial infections and these were thought to be due to toxins or toxic chemicals.

Evidence for the existence of a novel type of disease-causing organism or agent was first provided by work on an economically important disease of tobacco crops. In 1892, Iwanowski at St Petersburg discovered a new phenomenon by transmitting the tobacco mosaic disease to healthy plants by rubbing them with bacteria-free extracts of diseased leaves. Although this report went unnoticed until 1898, Loeffler & Frosch, studying foot-and-mouth disease, and Beijerinck, working with tobacco mosaic disease, confirmed that infectious agents could be serially transmitted from bacteria-free filtrates. Our concept of viruses arose from these observations since it was clear that these agents could not be simply toxins or poisons. For example, if a chemical, such as cyanide, is introduced into a 'victim', the effect will be quantitatively related to the amount of cyanide given and during the illness the amount of poison does not increase. On the other hand, new infectious material is necessary for the spread of a virus disease, such as smallpox, throughout the population which means that some type of multiplication of the causative agent must occur during the progress of the disease.

Multiplication is a principal phenomenon of living cells either in the increase of bacterial colonies or in the development of an organism. The intriguing fact about virology is that viruses are not cells. How then can they possess this basic characteristic of life?

1.2. The requirements for virus growth

Early definitions of viruses were based mainly on an attempt to differentiate them from other infectious agents such as bacteria. The essential difference was the observation that viruses, unlike other micro-organisms, could not multiply in chemically defined solutions. Viruses would only multiply in the presence of particular living cells or host organisms. Further, the bacteria-free extracts could often retain their infectivity for long periods, but they could not be cultivated in isolation from living tissue. Much later it was realised that viruses replicate only inside living cells making use of the internal

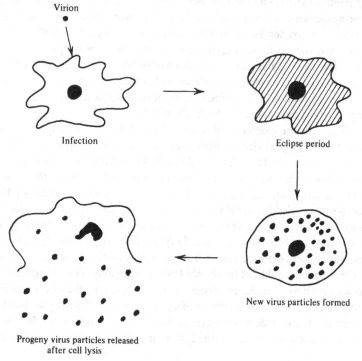

Virion

Infection   Eclipse period

New virus particles formed

Progeny virus particles released
after cell lysis

Fig. 1.1. The process of infection: a virus will only multiply within a living cell.

cellular environment. For purposes which will become clear later, this internal cellular environment, which is created by the immediate function of genes, will be referred to as the **genosphere**, in contrast to the **biosphere** which is the result of the interaction and function of living cells. As we shall see (fig. 1.1.), it is only within the sphere of sub-cellular organelles such as nuclei, mitochondria, ribosomes and cytoplasmic components that a virus can display the characteristics which justify calling it alive. Later in this book I shall develop the theme that Man lives in the biosphere whereas viruses live in the genosphere.

1.3. Isolation of tobacco mosaic virus (TMV)

In the early part of this century virology developed mainly as a clinical discipline within the framework of medical and veterinary sciences. Little progress was made towards determining the physical or chemical nature of viruses.

A major breakthrough came in 1935 when Stanley, an organic chemist, succeeded in purifying TMV. This virus multiplies in the leaves of tobacco plants and the cells become saturated with it. Kilogram quantities of infected leaves were collected and soluble extracts were prepared by grinding them up in aqueous solution and then removing the cellular material by filtration. At this time, there was some evidence that viruses may be proteins and by using procedures for the purification of protein, such as salt precipitation, Stanley obtained milligram quantities of purified TMV. Further, he demonstrated that the virus could be crystallised like a chemical molecule from concentrated solutions and that it retained its infectivity when re-dissolved. Here at last was a virus that could be obtained in a sufficiently pure state for accurate chemical analysis and it was soon shown that TMV contained nucleic acid as well as protein and behaved chemically as a **nucleoprotein**. These studies established the basic protocol of purification and analysis which allowed the subsequent development of virology as a quantitative scientific discipline. Stanley was awarded the Nobel Prize in 1946, the first of a long line of honours which have helped to bring virology to its present status.

1.4. The application of electron microscopy

By coincidence, the development of the electron microscope took place around the same time as the discovery of the molecular nature of viruses and the first electron micrographs demonstrated conclusively that these infectious agents were particulate. Although much smaller than bacterial cells, each particle appeared identical. TMV was seen to consist of a long rod, 15 nm in diameter and 300 nm in length, while other virus particles, such as turnip yellow mosaic virus (TYMV), were shown to be spherical in shape with a diameter of approximately 25 nm. For some years, plant viruses continued to contribute greatly to our knowledge of virus structure, mainly because they could be obtained in relatively large quantities. Furthermore, the development of the negative staining technique by the use of phosphotungstic acid for the rapid detection of virus particles by electron microscopy has played an important role in the discovery of new viruses and in the elucidation of their structures (fig. 1.2).

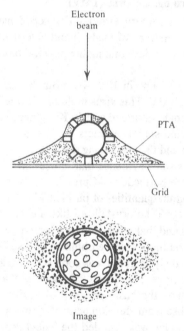

Fig. 1.2. A useful method of 'looking' at viruses is by negative-staining electron microscopy. PTA, phosphotungstic acid. (From Starke, G. & Hlinak, P. (1974). *Grundriss der Allgemeinen Virologie*. Veb Gustav Fischer Verlag, Jena.)

One of the most exciting discoveries made by the early electron microscopists concerned the viruses which infect bacterial cells. These viruses had been found as early as 1915 by Twort and also by d'Herelle in 1917 and were called **bacteriophages**, but since they did not appear to be associated with disease little attention had been paid to them. The interest generated by the first micrographs of these bacteriophages was immense; they obviously had a highly ordered and complex structure despite their very small size of approximately 100 nm. They had rounded heads and appendages which were initially thought to be tails, like minute tadpoles. After many years of controversy, it was established that the appendages had no motile action, but rather played an important role in the attachment of the phage particles to the surface of cells and also in the inoculation of the infectious viral nucleic acid.

1.5. The discovery of infectious phage DNA

The biological characteristics of viruses had shown that their behaviour could best be explained within the general context of genetics. They could mutate, propagate themselves and yet maintain highly defined characteristics such as host specificity, size and shape. Like other organisms they always, or nearly always, appeared to breed true. Direct chemical analysis showed that viruses contained nucleic acid as well as protein and a famous experiment by Hershey & Chase in 1952 established beyond doubt that the nucleic acid (DNA) of the bacteriophage was directly responsible for infection. The elegance of the Hershey & Chase experiment lay in the simplicity of approach and the fact that only a single, yet fundamentally important,

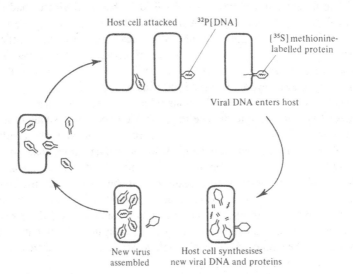

Fig. 1.3. Demonstration that the DNA of the bacteriophage is required for replication but the coat protein is not.

question was asked. What part of the virus enters the bacterium? By this time the use of radioactive isotopes was well established in biochemistry and it was possible to label nucleic acids with ^{32}P and protein with ^{35}S. Hershey & Chase made preparations of bacteriophage labelled with ^{32}P and ^{35}S, purified them and then used the labelled virus to infect non-radioactive bacterial cells (fig. 1.3). They found that the only radioactive element which entered the cells was ^{32}P. The proteins labelled with ^{35}S remained outside and could be removed from the bacteria without reducing the ability of the infected cells to produce progeny phage particles. The important implication here is that the replication of the phage is under the control of the nucleic acid whereas the protein plays no direct part in the transfer of information from one generation to the next.

Although this result might have been expected from earlier work by Avery, McCleod & McCarthy (1944) on the transformation of a non-virulent strain of bacterial cells by the addition of purified DNA from a virulent strain, the clear demonstration by the Hershey & Chase experiment left no doubt about the chemical nature of the genetic material. Their experiments opened the flood gates and led to our current knowledge of the biochemistry of genetics.

1.6. The development of mammalian tissue culture

These interesting discoveries in the field of bacterial viruses had mainly developed from the dedicated work of Max Delbruck and his collaborators in establishing the quantitative characteristics of bacteriophage replication. They stressed the ultimate importance of working with a simple, readily controlled system and realised the importance of what is now called 'the one step growth cycle', when all the cells of a population are undergoing synchronous or nearly synchronous infection. Only from such an experimental system can one deduce the probable events in a single cell. Also, of course, the yield of bacteriophages from cultures or bacteria was sufficient to allow accurate quantitative and chemical analysis to be done on them.

With animal viruses at that time no feasible way of studying the life cycle was available because of the complexity of the host organisms. Developments in this field continued to be predominantly of a clinical nature until around 1950. As is clear from the above comments on the growth of bacteriophages, a suitable experimental culture system was required and the answer to the animal virologists' problems arrived with the development of the technique of tissue culture.

Tissue culture is basically simple, and modern developments have made it one of the most widely used tools of virus research. In brief, small samples from an organ such as a kidney are removed and the tissue is dissociated into individual cells or minute pieces. If these fragments are put into a glass or plastic bottle in a suitable nutrient solution or medium, the cells will adhere to the surface of the glass and multiply to form cell sheets, generally called **monolayers** (see fig. 3.2, p. 23). Some cells, if kept agitated,

TABLE 1.1 *Composition of Eagle's Medium*

Component	Concentration (mg/l)
Inorganic salts	
Sodium chloride (NaCl)	6400.0
Potassium chloride (KCl)	400.0
Calcium chloride (CaCl$_2$)	200.0
Magnesium sulphate (MgSO$_4$. 7H$_2$O)	200.0
Sodium phosphate (NaH$_2$PO$_4$. 2H$_2$O)	140.0
Ferric nitrate (Fe(NO$_3$)$_3$. 9H$_2$O)	0.1
Amino acids	
L-arginine hydrochloride	42.0
L-cystine	18.0
L-histidine hydrochloride	19.2
DL-isoleucine	104.8
DL-leucine	104.8
L-lysine hydrochloride	73.1
DL-methionine	30.0
DL-phenylalanine	66.0
DL-threonine	95.2
DL-tryptophan	16.0
L-tyrosine	36.2
DL-valine	93.6
Vitamins	
Aneurine hydrochloride	2.0
Choline chloride	2.0
Folic acid	2.0
Inositol	3.5
Nicotinamide	2.0
Calcium pantothenate	2.0
Pyridoxal hydrochloride	2.0
Riboflavin	0.2
Miscellaneous	
Glucose	4500.0
L-Glutamine	292.0
Phenol red	15.0
Antibiotics	
Penicillin	200 000 units
Streptomycin	100 000 μg
Calf serum	5–10%

will grow in suspension just as bacterial cells do. In many cases, cells have become so adapted to growing under these conditions that they have been maintained for years in continuous culture with frequent renewal of the medium. Media are complex solutions (table 1.1) containing all the essential ingredients for supporting life, such as glucose, amino acids, vitamins, mineral salts and low concentrations of biological extracts such as serum. Although simple in concept, the practical problems involved in perfecting the technique to a routine procedure involved years of dedicated and tenacious work. The biggest hazard is the adventitious presence of bacteria or fungi. Since these grow more rapidly than animal cells, the slightest contamination soon destroys the culture.

Tissue culture was initially of great importance to the cancer researcher, but in the early 1950s it became firmly established that animal viruses from a large variety of diseases could be successfully grown in cultured cells. John Enders, who received the Nobel Prize in 1954 for his pioneering studies on the growth of poliovirus, provided a further landmark in the history of virology. The repercussions of the development of the growth of animal viruses in tissue culture still vibrate today with sustained vigour throughout virology. Tissue culture is currently used in diagnosis, vaccine production, the isolation of unknown agents and in fundamental studies on the replication and structure of viruses.

1.7. Crystallisation of poliomyelitis virus

Public recognition of the value of a novel development depends on a clear-cut demonstration of its potential. This was provided for tissue culture and virology by Schaffer & Schwerdt in 1955 when they successfully crystallised poliomyelitis virus. When a healthy monolayer of monkey kidney cells is treated with poliovirus, dramatic changes take place within a few hours; the cell die, fall off the glass and lyse. New virus is released into the medium. Although the yields of virus particles are only 50–100 μg/l of infectious fluid compared to bacteriophage yields of about 1 g/l, Schaffer & Schwerdt's experiments highlighted the potential value of the tissue-culture technique.

At about the same time, development of the method to an industrial scale was stimulated by the production of a vaccine against poliomyelitis. The Salk vaccine (see p. 131), first produced in 1953, was a triumph for both clinical virology and for the techniques of tissue culture. The success of the polio-vaccination programme gave great impetus to the development of vaccines against other important human and animal diseases. It should be remembered, however, that the basic principles of vaccination were established by Jenner in 1798 (see p. 130), a hundred years before the recognition of viruses as infectious agents. As we shall see later, many problems still persist which will only be resolved by a clearer understanding of the chemical basis of vaccination.

1.8. The modern era

The explosive growth of molecular biology during the last two decades mimics the logarithmic multiplication of viruses themselves and it would be difficult to single out any individual development from the stream of new information that has flooded the scientific journals. Here I shall mention only a few points, some of which will be treated more fully in later chapters. For example, it has been demonstrated that RNA as well as DNA could be the viral genetic material, the sequence of TMV protein has been elucidated and infective TMV has been reconstituted from its protein and nucleic acid components; the ultrastructure of many complex viruses has been determined. During the last ten years, great advances have been made in understanding the mechanisms of replication of virus nucleic acid and Haruna & Spiegelman in 1965 successfully obtained complete replication of the nucleic acid of a small RNA bacteriophage in a cell-free system. They showed conclusively that the viral nucleic acid was capable of self-generation provided that a suitable enzyme and precursors were present. Furthermore, the newly synthesised RNA was infectious and hence had retained its basic genetic characteristic of replicating and producing mature virus particles when injected into a suitable host cell. A few years later, in 1968, Goulian succeeded with a similar experiment to Spiegelman's on the DNA of another bacteriophage. These experiments illustrated that genetic material, both DNA and RNA, could be copied to yield biologically competent genes outside the living cell and the application of this development may well have far-reaching consequences. It is important to remember, however, that this 'test-tube' replication of nucleic acids is not a true chemical synthesis since it is completely dependent upon the presence of active enzymes which are direct gene products.

Probably the most rapidly advancing area of virology today is the study of cancer-inducing or **oncogenic viruses**. In 1911, Peyton Rous discovered an agent, now known as the Rous sarcoma virus (RSV), which could produce sarcomas (solid tumours) in chickens. Many viruses have since been found that are known to be the immediate cause of tumour formation especially in chickens and rodents. This area of virology has recently been recognised by the award of the Nobel Prize to three independent workers, Dulbecco, Temin and Baltimore, in 1975. Dulbecco and his colleagues had shown during the 1960s that certain small DNA viruses had alternative replication pathways. Usually these viruses replicate in and kill the host cell resulting in cell lysis. Occasionally, however, the infection does not lead to cell lysis but to cell transformation as a result of the stable integration of the virus genome into the host cells' chromosomes. On the other hand, the Rous sarcoma virus was shown to be an RNA virus and the way in which such viruses cause transformation of cells was a mystery until 1970 when Baltimore and Temin independently discovered a new type of enzyme called reverse transcriptase in RNA tumour viruses. Reverse transcriptase can make DNA from RNA molecules and hence also results in the stable integration of RNA virus genetic information into the host cell chromosomes. These discoveries have established that virology has an important role to play in the understanding of cancer as well as of the classical infectious diseases.

Of great current interest is the possiblity that some long-term degenerative diseases such as multiple sclerosis, schizophrenia, diabetes, rheumatoid arthritis and many other conditions may have a virus etiology and can result from complications arising from both virus infection and the immune system. During the 1950s a new brain disease called **Kuru** was recognised among the eastern highland cannibals of Stone Age New Guinea. This disease was a long-term degenerative condition and Gajdusek and his colleagues showed that it was passed on from generation to generation by the tribe's unusual eating habits. They also isolated from the affected people a virus agent, similar to the agent which causes a brain disease in sheep known as 'scrapie'. This group of viruses, often referred to as 'slow viruses', are receiving a great deal of attention as they may be involved with a host of major degenerative human diseases.

In summary, we see that virology has passed through three main phases. Firstly, it has been definitely proved that viruses are causative agents of many diseases. Secondly, the particulate and molecular nature of viruses has been demonstrated. Finally, the principal events of virus replication and a detailed outline of their chemical structure have been established. During this time, virology has matured into a scientific discipline that has not only a sound fundamental basis but also important medical, agricultural and perhaps sociological applications. From the quiet clinical search for the causative agents of disease during the early part of the century, we have recently passed through a more noisy, though exciting phase, illuminating the fundamental basis of biology and we can now approach the future from the firm foundation and understanding of the chemistry of viruses.

2 The principles of virus classification

2.1. The structural components

Viruses are divided into two major groups or sub-phyla based on the type of nucleic acid present. Virus particles contain only one molecule of nucleic acid; however, there are important exceptions to this rule and some viruses, such as influenza, contain a number of molecules of RNA, and in these the **genome** is said to be segmented. In all cases the nucleic acid is of one type, DNA or RNA, and is surrounded by a protein coat or shell. The protein coat is composed of a large number of **structural** units which consist of single polypeptide chains or a complex of different chains. Often these structural units are clustered in specific groups (**capsomeres**) which are usually large enough to be resolved in electron micrographs. A good example of this is seen in the common human wart virus (papilloma virus group) shown in fig. 2.1 where the visible morphological units are referred to as capsomeres.

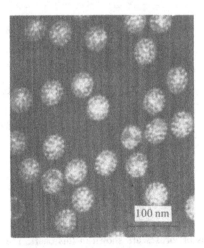

Fig. 2.1. Electron micrograph of a papilloma virus isolated from a human wart. (Photograph kindly supplied by Dr Elizabeth Hoey, Biochemistry Department, Queen's University, Belfast.)

A virus particle consists of an ordered complex of viral nucleic acid and structural protein sub-units. The entire structure is called the **nucleocapsid**. The term **capsid** is reserved for the protein shell which lacks the nucleic acid.

In some larger RNA viruses, the nucleocapsid is enshrouded by an **envelope** which is rich in glycoproteins and lipoproteins. These envelopes have many of the characteristics of cell membranes and are derived from the host-cell membrane during the final stages of virus **maturation**.

2.2. The size and shape of viruses

We have seen that the chief characteristic of viruses is their small size. They lie in the range between the largest protein molecules, such as haemocyanin (30 nm), and the smallest independent cells (300×500 nm). Fig. 2.2 illustrates the range of sizes and shapes of some common viruses. The crystallisation of TMV and poliovirus emphasised the molecular characteristics of viruses and there is clearly a case for considering them as macromolecules.

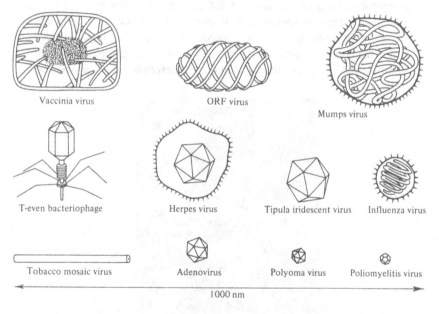

Fig. 2.2. Relative sizes of viruses are shown in this chart. The five viruses with polyhedral structures possess icosahedral symmetry. The tobacco mosaic virus and the internal components of influenza and mumps virus have helical symmetry. The remaining viruses exhibit complex symmetry. (From Horne, R. W. (1963). The structure of viruses. In *Readings from Scientific American, The Molecular Basis of Life*. W. H. Freeman & Co, San Francisco.)

It is very unlikely, however, that the larger viruses, especially those possessing envelopes, will ever be crystallised since there is evidence that both the shape and size of these particles can vary. It has been generally accepted therefore that the term 'particle' weight is used rather than molecular weight and that the latter is confined to the weight of the individual covalently bonded components.

One of the most striking features of viruses is the manner in which the structural units are arranged and this feature has been used as a major criterion for classification. Small RNA viruses and most of the DNA viruses have a spherical shape in which the protein units are arranged in the form of an icosahedron. On the other hand, the nucleocapsids of some of the RNA viruses and of many plant viruses are elongated and the sub-units are arranged in a helical fashion. The electron micrographs shown in fig. 2.3 illustrate the range of particle shapes that are found among some common viruses.

2.3. Virus classification

Problems of nomenclature and classification are often a stumbling block to students who wish to bridge the gap between the physical and biological disciplines. In this section the principles involved in a modern approach to classification of viruses will be outlined. Historically, many viruses were named after the description of the disease they caused. Hence the agent causing foot-and-mouth disease was called foot-and-mouth disease virus (FMDV). Similarly, TMV is a direct description of the pathogenic effect that the virus has on tobacco plants. The naming of many other plant viruses has followed this principle: sugar beet curly top virus (SBCTV), cranberry false blossom virus (CFBV) and potato witch's broom virus (PWBV). Traditional or trivial names of human diseases, followed by the word virus, are commonly used, such as measles, mumps, smallpox, influenza and rabies. Still others have been named after their discoverer, for example Rous sarcoma virus (RSV), or locality of occurrence or discovery, such as Fiji disease, Semliki Forest disease, Newcastle disease.

However, the student should not let this apparent state of anarchy deter him, since the situation has arisen mainly because the true nature of viruses has been defined only relatively recently.

The chief obstacle in classifying viruses has been the assumption that viruses are organisms and various attempts have been made to classify them on the Linnean binomial system analogous to the schemes used with plants and animals. Here, however, the purpose of classification is to group together into meaningful categories the organisms that are closely related.

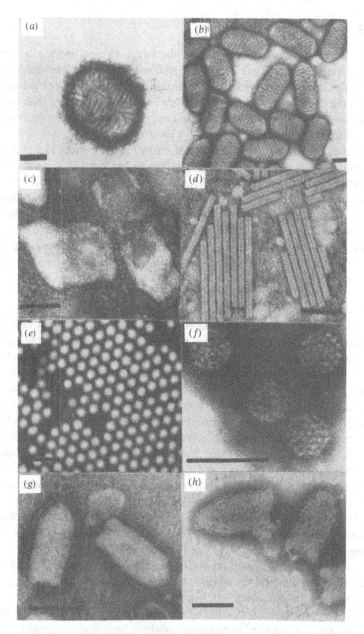

Fig. 2.3. Electron micrographs of some common viruses. Each bar line represents 100 nm. (a) Vaccinia (small-pox virus); (b) orf virus (paravaccinia); (c) measles virus; (d) TMV; (e) bovine enterovirus (a virus similar to polio virus and foot-and-mouth disease (FMD) virus); (f) papovavirus (human wart); (g) vesicular stomatitis virus (VSV); (h) rabies virus. Photographs were kindly supplied by Evelyn Dermott (a, b, c, f); P. Cooper (d); Elizabeth Hoey (e); C. Smale (g, h).

Although Linnaeus developed this method of classification about 120 years before Darwin formulated the theory of evolution, the scheme does reveal evolutionary and phylogenetic relationships. At present, there is no reason to believe that viruses form a single group of organisms, all members of which have a common ancestor, and virologists have had to find another framework on which to develop a logical and scientifically sound classification.

In contrast, the classification and nomenclature of chemicals has been based generally on structure. Elements are classified according to electronic configuration in the Periodic Table. At a slightly higher level of chemical complexity, carbon compounds are classified and named on the basis of structural and functional groups.

The discovery that viruses are particulate introduced the possibility of classifying them on their chemical, physical and structural characteristics rather than on their biological properties alone. The International Committee for the Nomenclature of Viruses favours a scheme in which all viruses with similar chemical and structural properties are grouped together regardless of whether they infect plants, animals or bacteria. They also suggest that a Latinised binomial system be used in accordance with the systematics of

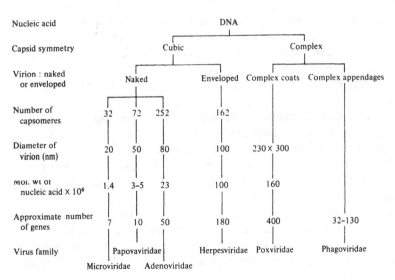

Fig. 2.4. Classification of DNA viruses.

plants and animals, making it possible to designate a given virus by its genus and species name. In this classification, all viruses are placed in a single phylum and DNA and RNA viruses each constitute a sub-phylum. The remaining sub-divisions, orders, families and genera, depend on the presence or absence of an envelope and the shape of the nucleoprotein component. The principles underlying the classification scheme are outlined in fig. 2.4 for DNA and in fig. 2.5 for RNA.

2.4. The cryptogram

For many purposes, the most straightforward approach to classification has been the system devised by Gibbs, Harrison, Watson & Wildy in 1966 and modified by Wildy in 1971. They suggested that viruses would automatically fall into groups only after a systematic analysis of the chief chemical, physical and morphological characteristics had been made. The information could be assembled in **cryptograms** which would be similar for similar viruses. The characteristics of any virus can be defined by a cryptogram which contains at least four sets of terms, such as

$$\frac{a}{b} : \frac{c}{d} : \frac{e}{f} : \frac{g}{h}.$$

Each term is an alphanumeric code relating to a particular property of a virus particle. The great value of this approach is that a cryptogram summarises the essential characteristics in an easily comprehensible form. Each letter in the cryptogram is defined as follows:

(a) describes the type of nucleic acid: D is written for DNA viruses; R for RNA viruses;

(b) describes whether the nucleic acid is double (2) or single (1) stranded;

(c) refers to the molecular weight of the nucleic acid (in millions);

(d) refers to the percentage nucleic acid in the virion;

(e) describes the general shape of the virion: S = spherical, E = elongated with parallel sides but end not rounded, U = elongated with parallel sides with rounded ends, X = complex;

(f) describes the shape of the nucleocapsid as in (e);

(g) describes the host organism: V = vertebrate, I = invertebrate;

(h) describes how the virus is transmitted: 0 means that no vector is required, Ac, Si, symbols referring to insect vectors such as mites or fleas.

When a characteristic of a virus is unknown an asterisk is inserted in the appropriate position in the cryptogram. This procedure has obvious value in emphasising the state of our ignorance about some viruses. On the other

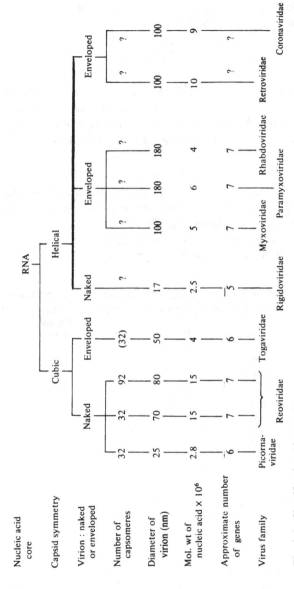

Fig. 2.5. Classification of RNA viruses.

hand, our knowledge of some viruses is very precise and poliovirus can be fully described by the cryptogram

$$\frac{R}{1} : \frac{2.6}{30} : \frac{S}{S} : \frac{V}{0}.$$

This means that polio is a single-stranded RNA virus. The RNA has a molecular weight of 2.6×10^6 and it constitutes 30% of the mass of the virus particle. The particle is spherical in shape, has no envelope and it infects vertebrates. In contrast, the cryptogram for scrapie virus is

$$\frac{*}{*} : \frac{*}{*} : \frac{*}{*} : \frac{V}{0}.$$

A simplified and abbreviated form of virus classification is included as the Appendix. This includes the cryptograms of each group and the names of most of the viruses mentioned in this book.

The reader should now examine table 2.1 and for practice decode the cryptograms in consultation with the Appendix.

TABLE 2.1. *Complete the following table by reference to the Appendix*

	Cryptogram	Family	Genus	Possible disease
	$\frac{a}{b} : \frac{c}{d} : \frac{e}{f} : \frac{g}{h}$			
1	$\frac{D}{2} : \frac{130}{15} : \frac{S}{*} : \frac{I}{*}$			
2	$\frac{R}{1} : \frac{7}{1} : \frac{S}{E} : \frac{V}{0}$			
3	$\frac{D}{2} : \frac{200}{6} : \frac{X}{X} : \frac{V}{0}$			
4	$\frac{R}{1} : \frac{\Sigma 10}{2} : \frac{X}{X} : \frac{V}{0}$			
5	$\frac{R}{2} : \frac{\Sigma 15}{15} : \frac{S}{S} : \frac{V}{0}$			
6	$\frac{D}{2} : \frac{4}{10} : \frac{S}{S} : \frac{V}{0}$			

2.5. Variety and specificity

Even within an individual virus family there is a fascinating range of characteristics and properties. For example, the sub-family Picornavirinae contains six genera (see Appendix). Representative viruses from all these genera are very similar in morphological and chemical properties and yet the diseases they cause are strikingly different ranging from the common cold (rhinoviruses) and poliomyelitis in man to foot-and-mouth disease in cattle. All cellular organisms, from the smallest bacterial cell like *Escherichia coli* to the largest mammals, appear to act as hosts to specific viruses which apparently show little or no affinity for other species. These observations alone help us to appreciate the problems of classification and justify early attempts to classify viruses from an **epidemiological** viewpoint.

The essential feature of all virus diseases is the ability of the invading virus to infect cells in the host organism. The nature of this specificity is intriguing and at present we can only explain it in an imprecise manner; but current developments in our knowledge of the chemical structure of viruses and the composition of cell membranes are indicating that cell surfaces contain specific receptor sites which can bind complementary sites on the virus surface. This highly specific attachment must take place before the process of infection can be initiated.

2.6. The antigenic nature of viruses

There are generally two consequences when a virus infects a warm-blooded organism. First, the virus enters a cell and multiplies, producing a large number of new virus particles which are often released into the blood stream and infect other cells. Secondly, the infected organisms react against this invasion by producing **antibodies** which can bind to and inactivate the virus. The antibodies combine with the **antigens** which form part of the surface of the virus particles. The exact chemical nature of specific **antigenic sites** is not known at present, although it is highly probable that unique amino acid sequences in the coat proteins or possibly the carbohydrate components of the enveloped viruses are involved.

Very minor changes in the sequences of coat proteins can result in the formation of different sub-types of a virus. Generally, antibodies produced against one sub-type will not react or **neutralise** other sub-types. This forms the basis of a serological neutralisation test which makes it possible to differentiate between different sub-types. This serological classification is of the utmost importance from a practical viewpoint since it is the method used to diagnose viral disease. It also provides the most definitive method of classifying any particular virus and the aim of many biochemical virologists is to understand the chemical differences between serotypes. There are, for example, three known serotypes of poliovirus. There are seven major serological types of FMDV and within each main group there are a number of sub-types which can differ sufficiently to affect the success of vaccination programmes. Here we see a classic example of the problem facing epidemiologists when any one of more than 60 different sub-types may be causing the outbreak of foot-and-mouth disease.

3 Quantitative methods of assaying viruses

3.1. Principles of virus titrations

The quantitative determination of virus suspensions makes use of two characteristics which are common to all viruses. The first method, an **infectivity** assay, depends on the ability of viruses to multiply, generally at a very rapid rate, within the host cell and subsequently to release infective particles. This usually results in the infection of neighbouring cells and leads to the spread of infection throughout the organism or cell culture. In many ways the process is rather analogous to the amplification that occurs in a photomultiplier. The virologist allows this 'biological amplification' to continue until the result is visibly obvious. The second method, an **immunological** assay, does not depend on the ability of viruses to cause infection, but rather on the specific nature of their surface components. As already mentioned, all viruses can behave as antigens and, once a specific antiserum has been prepared, standard immunological procedures can be used for determination of viral-specific components. It is important to remember that an infectivity assay only tells us about the number of infective units present and provides no information concerning non-infective particles. On the other hand, an immunological assay generally refers to the amount of a specific component present regardless of whether or not the sample contains infective particles.

3.2. Plaque assay

The **plaque assay** procedure has become firmly established as the most accurate method for determining the number of infective units present in a virus suspension. It was initially used for titrating bacteriophages but the method has become the mainstay of many animal virus laboratories since the development of mammalian cell cultures. The technique involves the infection of a number of monolayers of cells with the virus suspensions over a range of high dilutions. After a short period the infected cell sheet is covered with a layer of liquid nutrient agar, which rapidly solidifies and prevents the spread of the products of virus multiplication throughout the medium. Secondary infections are therefore localised in a small area of cells close to the site of primary infection. After appropriate periods which depend on the type of virus, the regions of dead cells can be visualised by staining procedures as shown in fig. 3.1. It has been shown that at high dilutions a single plaque is initiated by only one infective particle and hence from the number of plaques formed at a given dilution it is possible to calculate the total number of plaque-forming-units in the original virus suspension. The results are generally expressed as 10^x plaque-forming-units (p.f.u.)/ml, but

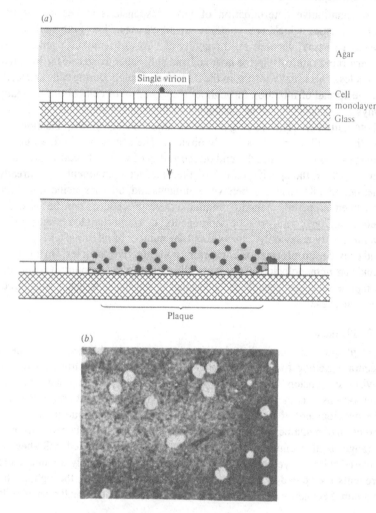

Fig. 3.1. Plaque assay. (*a*) The process of plaque formation. (*b*) Photograph of plaques produced by bovine enterovirus on baby hamster kidney cells (BHK).

it must be emphasised that this refers only to the number of infective particles and not to the numerous non-infective particles which are generally present in virus suspensions.

3.3. Quantal assay methods

Other methods are available which also determine the concentration of infective particles but do not depend on the direct enumeration of infective centres or plaques. The assays depend on unmistakable observations, such

(a)

(b)

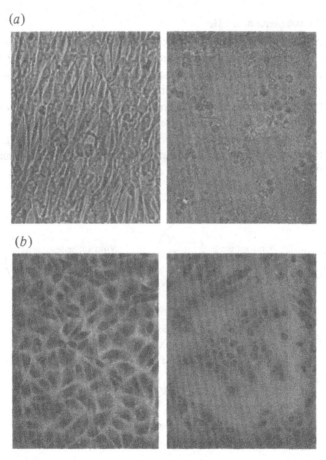

Fig. 3.2. Typical cytopathic effects produced by animal viruses. (a) Bovine entero-virus (a picornavirus) causes rapid cell lysis and destruction of the monolayer. (b) Measles virus (a paramyxovirus) causes formation of large multi-nucleated cells called syncytia. The virus causes neighbouring cells to fuse together. Uninfected cells, left frames; infected cells, right frames.

as the death of a mouse or severe **cytopathic effect** in a culture of cells (fig. 3.2). Replicate groups, each containing at least five individuals, are inoculated with a small volume of virus suspension at an appropriate dilution. The animals or cultures are observed daily and an end-point is determined by finding the dilution which causes half the individuals of a group to show a positive and half to show a negative effect. The results are expressed as Tissue Culture Dose (TCD50) or Lethal Dose (LD50). Although occasionally less accurate than the plaque assay these quantal methods are technically simpler and are often used with viruses which do not readily form plaques.

3.4. The complement fixation test

Immunological methods of assay depend on the fact that viruses can cause animals to produce specific antibodies. Antisera produced against known viruses can therefore be used to established the presence and nature of unknown agents and are especially important in clinical diagnosis. A large variety of tests have been devised which make use of antisera, but only the principal ones will be described here. The **complement fixation test** is one of the most accurate methods of determining the amount of antigen present. The procedure depends on the fixation of complement during an antigen–antibody reaction. Complement is a complex of proteins conveniently ob-

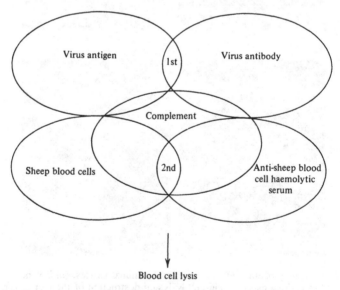

Fig. 3.3. Complement fixation test. (For explanation see p. 25.)

tained from guinea-pig serum. The test consists of two experimental phases. First, antibody and antigen are mixed with a standard amount of complement and incubated for an hour at 37 °C. The amount of residual or non-fixed complement is determined by a second test which uses sheep blood cells that have been sensitised with anti-sheep cell haemolytic serum. The sensitised sheep-blood cells are added to the initial mixture and if free complement is present they will be lysed. On the other hand, if all the complement has been used up or 'fixed' in the first antigen (virus) – antibody reaction, there will be no lysis (fig. 3.3). The assay can be made highly quantitative by determining the amount of lysis that occurs by spectrophotometric analysis of the completed assay. The results are generally expressed as units of complement fixed (CF units), and are therefore directly related to the amount of antigen present. It must be remembered, however, that these values are relative and cannot be expressed in absolute units such as micrograms or milligrams.

3.5. The neutralisation test

This is probably the most specific way to test for virus antibody. It is generally type-specific and is widely used in clinical laboratories. It forms the basis of the serological classification of viruses as described in § 2.6 and is the most highly specific virological test yet devised. Although there are numerous variations of the procedure, the general principle is quite simple. Dilutions of antiserum are mixed with a given concentration of virus and allowed to react for a period depending on the particular virus being studied. Samples of the mixture are then assayed for virus infectivity by one of the methods described in § 3.2 or § 3.3. Clearly, only antiserum which is specific for the virus will neutralise the infectivity, but the technique can also be used to establish the occurrence of **cross-neutralisation** between closely related sub-types.

3.6. Immunofluorescence

The sensitivity of immunological procedures has been greatly increased by the development of techniques of tagging antibodies with **fluorescent** dyes which fluoresce on exposure to ultraviolet light. Fluorescent antibodies can be readily prepared in the laboratory and are commercially available. This technique gives the ability to detect virus-specific antigens inside cells and hence the locality of virus antigens can be traced. Two main procedures are used. Direct immunofluorescence utilises virus-specific antibody which is tagged. Indirect immunofluorescence, sometimes called the 'sandwich' method, differs in that the antiviral antibody is not tagged but is treated with fluorescent antigamma globulin after it is fixed on to the cell. The indirect technique has the advantage of greater sensitivity and especially in diagnostic virology requires only a single tagged reagent. Fig. 3.4 shows the antigens present in cells persistently infected with measles virus.

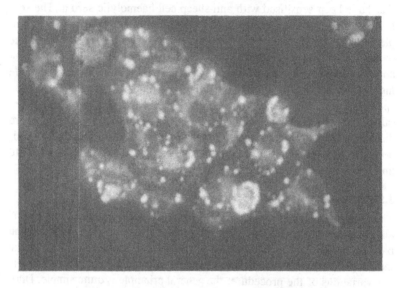

Fig. 3.4. Immunofluorescence test, Small monolayers of measles-infected cells were fixed with acetone and then covered with antiserum. After incubation the non-bound serum was washed off and the cells covered with anti-human γ-globulin which had been labelled with isothiocyanate. The bound fluorescent dye can be seen as bright (yellow/green) spots in a microscope equipped to use ultraviolet light. The bright spots in the photograph above indicate the locality of virus antigens.

3.7. Radio-immunoassay

Radioactive iodine (^{125}I) can be easily coupled to the tyrosine residues of proteins and this property has been used in the development of the very sensitive procedure of **radio-immunoassay**. The method of auto-radiography of cells labelled with [^{125}I]antigen or antibody is over a thousand times more sensitive than fluorescent procedures. The procedure can also be used for the quantitative estimation of specific antibody or antigen in solutions by use of affinity chromatography or precipitation of antigen–antibody complexes, and the bound radioactivity is determined by counting in a scintillation spectrometer.

3.8. Enzyme-linked immunoassays

More recently, application of the concept of tagging readily detectable components to antigens or antibodies has been extended to the use of enzymes. For example, the enzyme alkaline phosphatase or peroxidase can be covalently linked to immunoglobulin and the presence of fixed antibody–enzyme can be detected by colorimetric means. This method, although novel, is of great potential in the rapid diagnosis of virus diseases as the test can be carried out on filter discs which will turn a particular colour (depending on the enzyme assay used) whenever specific virus antigens are present. A possible procedure is illustrated in fig. 3.5 and although probably

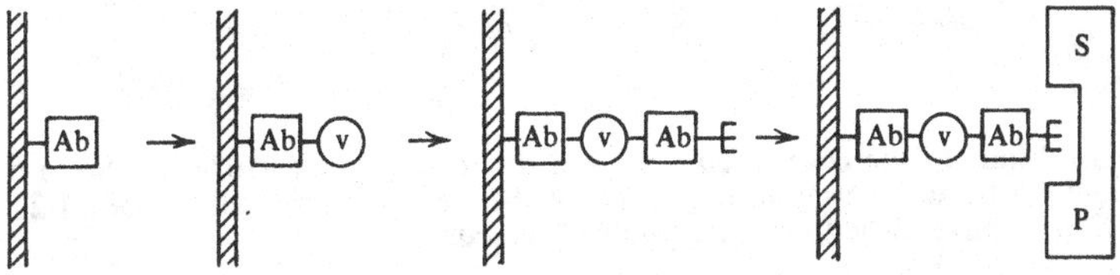

Fig. 3.5. Enzyme-linked assay. This test is still being developed but it should provide a sensitive colorimetric method for rapid viral diagnosis. In the scheme illustrated the hatched areas represent a solid matrix such as paper, plastic, Sepharose or glass. Ab, antibody; V, virus; E, enzyme such as alkaline phosphatase; S, substrate such as *p*-nitrophenyl phosphate; P, product such as *p*-nitrophenol which has a yellow colour.

not as sensitive as radio-immunoassay procedures the simplicity of the test lends itself to the rapid diagnosis of virus infections outside the laboratory especially in farming areas and in underdeveloped countries.

3.9. Immuno-diffusion

Antigen–antibody interactions often result in precipitation of the complexes and this has been utilised in the **gel diffusion test**. The technique involves the use of agar gels in which antibodies and antigens diffuse towards each other and when they meet they form a sharp line of precipitate. The method is particularly valuable in detecting the presence of multiple antigenic components of viruses.

	2	4	8	16	32	64	128	256	512	1024

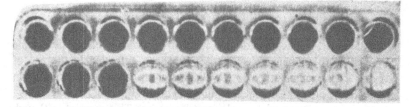

(*a*) Zero time. The cells have not settled in any of the cups.

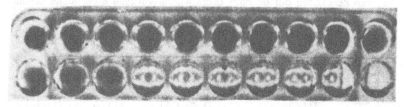

(*b*) One hour. In the control cups, and in the most dilute (1/1024) of the upper row, the cells are settling into a button in the centre of the cup. From 1/2 to 1/512 they have settled over the whole of the cup.

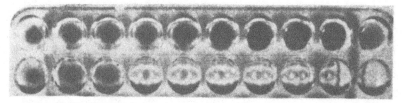

(*c*) Six hours. In those cups containing the most virus the enzymic action of the virus has freed it from the cells, which are sinking to the centres of the cups.

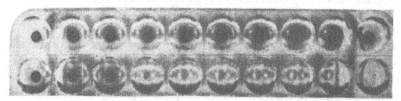

(*d*) Twenty-four hours. Elution of virus from cells has proceeded in all the cups containing virus, which now look very like the three control cups.

Fig. 3.6. Haemagglutination test. Haemagglutination of red blood cells by Newcastle disease virus, and subsequent elution of virus from the cells. The pictures are of the same plastic plate at various intervals after addition of a suspension of fowl red cells. The cups in the top row contain serial dilutions of allantoic fluid from an infected egg, from 1/2 to 1/1024. The three cups in the lower row contain saline. The test was carried out at room temperature. (From Waterson, A. P. (1968), *Introduction to Virology*. Cambridge University Press, Cambridge & London.)

3.10. Haemagglutination and haemolysin tests

Many viruses are absorbed by and cause the association of red blood cells of various species of animals and birds. This property, known as **haemagglutination**, provides a very rapid method for determining the presence of virus in a suspension. The reaction is generally carried out on plastic plates composed of rows of small cups. Fresh erythrocytes and dilutions of the virus suspension are added to the cups and left for 30–45 min. In control cups, where virus is not present, the erythrocytes sink to the bottom as a compact pellet, whereas if virus is present and agglutination occurs the erythrocytes remain as a diffuse layer distributed over a much larger area (fig. 3.6). In the Myxovirus group the process is reversible since the virus contains an enzyme capable of dissociating the aggregate.

Some viruses can cause the lysis of red blood cells and this can also be used as a rapid method of detection called the **haemolysin test**.

Although these procedures are much less sensitive than titration for infectivity the advantages of a rapid result from such a simple technique are considerable.

4 Purification of viruses

4.1. Preparation of virus harvests

Many of the biological characteristics of viruses can be determined on preparations which contain substantial amounts of host-cell debris and products. Analysis of properties such as infectivity, haemagglutination, serotype and virulence can be done directly on fluids isolated from infected tissue. For chemical analysis, however, it is essential to isolate the virus particles in a pure state. As we have seen, bacterial and animal-cell cultures are convenient sources of susceptible cells and initial concentration of virus in the range 10^5–10^{10} p.f.u./ml can often be obtained. With plant viruses, the intact plant can be a useful reservoir of virus. Often, the leaves retain the newly formed virus particles and a suitable harvest can be obtained by grinding or homogenising them in aqueous buffers. In contrast, the use of intact animals for the preparation of pure virus is of little value, since the problems of purification are considerably increased. Some animal viruses grow well in hens' eggs and these are used extensively in the study of influenza virus. In general, a virus harvest consists of a suspension of cell-breakdown products, virus particles and other virus products. The essential problem of purification is to remove all the host-cell material without damaging or inactivating the virus. The problem is immense in quantitative terms, since the amount of cellular material is often many thousands of times greater than the actual mass of virus particles. Also, the chemical natures of both the cell products and virus particles are very similar.

Purification schemes generally depend on two main characteristics. First, viruses behave as though they were very large proteins and therefore the techniques used for the purification of proteins are of value. Secondly, viruses often possess a highly defined size, shape and density and therefore techniques which can effect a separation on the basis of these properties are important.

4.2. Differential centrifugation

Viruses are smaller than most sub-cellular organelles and the first step in purification is generally centrifugation in order to remove the bulk of cellular debris, including nuclei, mitochondria and cell membrane fragments. Some viruses such as TMV can be highly purified by repeated cycles of slow (5000–10000 r.p.m.) and high (50000–60000 r.p.m.) speed centrifugation. For most animal viruses, however, a combination of several methods is required before an acceptable degree of purity is achieved.

4.3. Precipitation procedures

'Salting out' by ammonium sulphate has long been used in the isolation of proteins and enzymes and has had wide application in the initial stages of virus purification. The production of even milligram quantities of most viruses usually involves an initial volume of a few litres of virus harvest and precipitation procedures can readily effect a rapid means of concentration. Recently, zinc acetate has become a popular precipitant; at neutral pH, a flocculent precipitate of zinc hydroxide forms which efficiently co-precipitates virus and other protein present. This precipitate readily dissolves in a buffer containing ethylenediamine-*NNN'N'-tetra*-acetic acid. The technique has been successfully applied as a concentration step in the purification of rabies virus and enteroviruses.

4.4. Gel filtration

After concentration of the virus suspension to a convenient volume, the next steps depend on more refined techniques for separating molecules of different sizes and shapes. The technique of gel filtration or exclusion chromatography has been used extensively in the separation of many virus particles. The commercially available 'Sepharose' gels have well defined pore sizes which only allow molecules below a certain size to penetrate the interior of the gel beads. Large particles are excluded and therefore pass rapidly through a column of gel. A wide range of gels is available which allows separation of particles in the molecular weight range 10^2–10^7. 'Sepharose' 2B, which has an exclusion limit of 20×10^6 daltons, is suitable for viral studies. Many viruses are not completely excluded from 'Sepharose' gels and hence gel filtration can be used for estimating the relative sizes of different virus particles. When gel filtration is used in purification procedures, it is advantageous to select a grade of gel which will not completely exclude the virus particle. In this way, it is possible to separate the virus from any larger or smaller particles which are present. On the other hand, if a gel type is selected from which the virus is completely excluded, then no separation from contaminating larger particles will be effected. Gel filtration has been used in the purification of a bovine enterovirus and fig.

4.1 shows how the virus can be separated from the bulk of other cellular contaminants. It must be remembered, however, that gel filtration does not separate virus from other particles of similar dimensions.

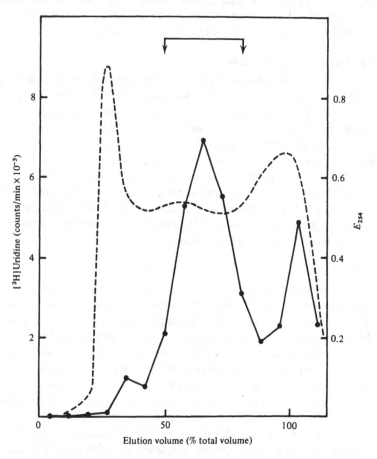

Fig. 4.1. Gel filtration of virus concentrate through Sepharose 2B. ----, E_{254}; ●——●, ^{3}H counts/min. (from Martin, S. J., Johnston, M. D. & Clements, J. B. (1970). *J. gen. Virol.* **7**, 103–13.)

4.5. Sucrose-gradient sedimentation

This technique involves sedimentation of a sample through a solution of sucrose whose viscosity, density and concentration increases in a linear fashion from the top to the bottom of the centrifuge tube. Centrifugation

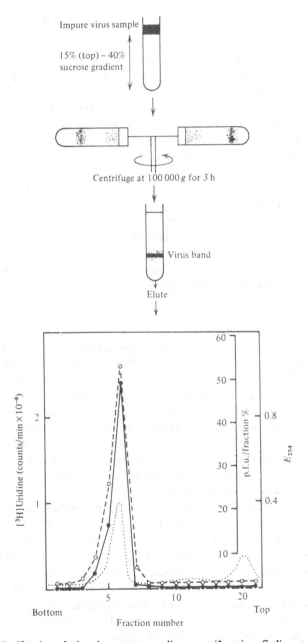

Fig. 4.2. Purification of virus by sucrose-gradient centrifugation. Sedimentation of bovine enterovirus through a 15–40% sucrose gradient. ·····, E_{254}; ●——●, counts/ min.; O----O, infectivity, p.f.u./fraction %. (From Martin, S. J., Johnston, M. D. & Clements, J. B. (1970). *J. gen. Virol.* **7**, 103–13.)

is generally carried out at approximately 30000–40000 r.p.m. and the gradient is then fractionated into small volumes by elution via a small hole pierced in the bottom of the tube as shown in fig. 4.2. If reference particles of known sedimentation coefficients (S) are available, fairly accurate estimations of the S values of unknown particles can be made. For many purposes, ribosomes are a useful source of reference particles; they have a sedimentation coefficient of 80 S. Viruses have sedimentation coefficients between 120 S and 1000 S except for a few small viruses like brome grass mosaic virus (86 S), broad bean mottle virus (84 S) and the bacteriophages such as $Q\beta$ (84 S) and f2 phage (80 S). The sedimentation rate of particles depends not only on their molecular weights, but also on their shape and density and hence the technique is extremely valuable in separating particles which have the same or similar sizes. Many viruses produce particles that do not contain the nucleic acid molecule but are the same size as the infectious virus. These non-infectious 'empty' particles can be easily separated from the virus particles by sucrose-gradient sedimentation (see §5.14) since they have vastly different S values.

4.6. Isopycnic centrifugation

Solutions of heavy metal salts such as caesium chloride or caesium sulphate will form linear density gradients when centrifuged for long periods at high speeds. Since the exact nature of the gradients depends on the initial concentration of the salt and the rate of centrifugation the range of densities that can be established may be precisely controlled. In practice, the gradients are generated in small plastic tubes and centrifugation is continued for over 20 h at approximately 60000 r.p.m. Immediately after centrifugation a hole is pierced in the bottom of the tube and the contents are collected in small volumes. The densities of the fractions can be determined by measuring the refractive index with a refractometer.

If particles of different densities are suspended initially in the solution, they will migrate and concentrate during centrifugation at positions where their density equals that of the gradient. Viruses possess characteristic densities and the method is particularly useful for separating them from cellular contaminants. Simple viruses, such as picornaviruses which contain only nucleic acid and protein, generally have a density in the range 1.33–1.45 g/cm^3 as shown in fig. 4.3. Viruses which possess an envelope have much lower densities in the region of 1.20 g/cm^3. For the latter type of viruses, potassium tartrate solution has become a popular medium for centrifugation, as well as caesium chloride, since it does not cause breakdown of these more fragile viruses. The lipid-containing viruses can often form visible bands in

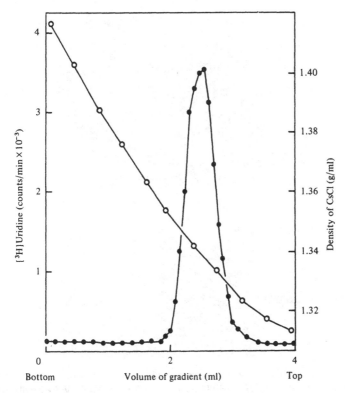

Fig. 4.3. Purification of bovine enterovirus virus by centrifugation through caesium chloride gradients. ●——●, ³H counts/min.; O——O, density CsCl g/ml. (From Martin, S. J., Johnston, M. D. & Clements, J. B. (1970). *J. gen. Virol.* **7**, 103–13.)

density gradients even when only very small quantities of material are present, as shown in fig. 4.4.

4.7. Extraction by non-aqueous solvents

As mentioned in §4.1, purification methods must not inactivate the virus particles. Consequently, organic solvents are of little value in the purification of viruses containing lipid, but are widely used with simpler viruses, such as poliovirus and foot-and-mouth disease virus. The most popular solvents are butanol and tri-fluoro-tri-chlorethane. The organic solvent not only removes the host-cell lipid components, but also denatures much of the host-cell protein which then collects as an aggregate at the interphase.

Fig. 4.4. Purification of measles virus by sedimentation in potassium tartrate gradients. Due to the enveloped nature of the virus a white band can be readily seen.

4.8. Treatment with detergents and enzymes

Water-soluble detergents, such as sodium deoxycholate, sodium dodecyl sulphate, and Tween (fig. 4.5) have been used extensively in the purification of many types of viruses. They are particularly useful in the study of complex viruses (see p. 71) when selective removal of the lipid components can be effected, leaving the nucleocapsids intact. It must be established that detergents do not inactivate non-enveloped viruses and that associated labile virus particles which may be of interest are not lost.

Purification can often be aided by treatment with deoxyribonuclease, ribonuclease, trypsin, pronase or papain. These enzymes may selectively degrade cellular nucleic acids and proteins without damaging the virus particles. Although the infectivity of some viruses is resistant to such enzyme treatment, it is always difficult to prove that 'minor' structural features have not been altered.

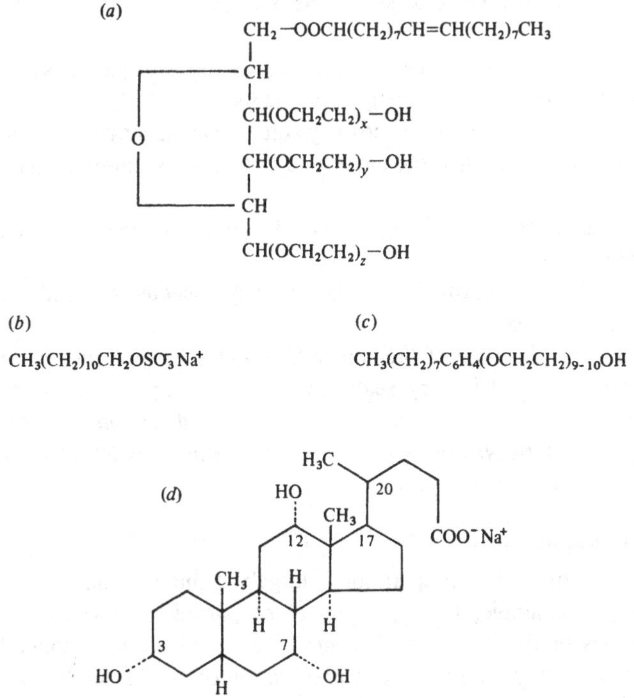

(a)

$$CH_2-OOCH(CH_2)_7CH=CH(CH_2)_7CH_3$$
$$|$$
$$CH$$
$$|$$
$$CH(OCH_2CH_2)_x-OH$$
$$|$$
$$O \qquad CH(OCH_2CH_2)_y-OH$$
$$|$$
$$CH$$
$$|$$
$$CH(OCH_2CH_2)_z-OH$$

(b)

$$CH_3(CH_2)_{10}CH_2OSO_3^-Na^+$$

(c)

$$CH_3(CH_2)_7C_6H_4(OCH_2CH_2)_{9-10}OH$$

(d)

Fig. 4.5. Chemical formulae of some common detergents used in the purification of viruses. (a) Tween 80 (polyethoxy sorbitan mono-oleate); (b) Sodium dodecyl (lauryl) sulphate, an alkyl hydrogen sulphate; (c) Triton X-100, an alkyl aryl polyethoxy alcohol; (d) Sodium deoxycholate, a steroid.

4.9. A purification protocol

It is seldom, if every, necessary or advisable to make use of all the available methods in the purification of a virus. The methods chosen will depend greatly on the individual characteristics of the virus being studied and on the particular requirements of the experiment. Although it is impossible to give a standard scheme for the purification of viruses, a few guide-lines can be formulated which typify the step-wise procedures now in common use.

Step 1. The virus harvest is centrifuged at low speed to remove macroscopic cellular debris.

Step 2. The virus and remaining contaminants can be concentrated by precipitation procedures.

Step 3. After resuspension of the precipitate in a suitably small volume of buffer an impure virus pellet can be obtained by centrifugation at high speeds.

Step 4. Gel filtration can be used to separate the virus particles from both larger and smaller contaminating components.

Step 5. Sedimentation on sucrose gradients can separate the virus from particles which may have the same size but possess different shapes and densities.

Step 6. Isopycnic centrifugation can be used to separate particles of different densities.

Step 7. The final 'purified' virus is generally collected as a pellet by high speed centrifugation.

Step 8. Crystallisation of the concentrated virus suspension would be an ideal final stage, but is very seldom carried out in practice except when required for specific purposes such as X-ray diffraction analysis. It is unlikely that many viruses will ever be crystallised owing to their pleomorphic and enveloped nature.

4.10. Criteria for purity

The 'purity' of a virus preparation is largely defined by empirical criteria such as, for example, the homogeneity of particles as seen in electron micrographs or the presence of single homogeneous peaks isolated from velocity or density gradients. The preparation of an absolutely pure virus suspension is probably unattainable for a number of reasons. For example, in any population of virus particles only some are infectious even though they all appear to be physically similar. Again, usually more than one type of particle, such as precursors and by-products, is produced during virus multiplication and these impurities must be removed during purification procedures.

Hence purification procedures must perform two main functions. First, they must remove the host-cell components which are present in the initial cell lysates; secondly, they must separate the viral components which differ in infectivity or chemical properties.

The successful removal of host-cell components can be monitored by a technique involving the use of non-infected cells which have been grown in the presence of radioactive precursors such as $^{32}PO_4$, [^{3}H]uridine or [^{14}C]amino acids. In a typical experiment a homogenised suspension of these radioactive cells is added to a non-radioactive virus lysate and the virus is then purified. The presence of host-cell material can be detected readily by measuring the radioactivity and the purification scheme may be considered successful if the final 'purified' virus is not radioactive. Although this method has been used successfully for picornaviruses, it is not satisfactory for enveloped viruses. As these viruses mature by a **budding process** it is possible that they trap cellular material into the virus structure. A method of testing whether this happens is to label cells with [^{35}S]methionine prior to infection and to allow the virus growth to take place in the presence of a large excess of the non-radioactive precursor. Any radioactivity detected in purified virus will indicate that some host-cell protein has been incorporated into the virus preparation.

5 Architecture of viruses

5.1. The genetic economy of multi-component systems

A basic characteristic of all genetic material is that the nucleic acid is always protected from the extracellular environment by a membrane or coat; in viruses this is achieved by the capsid. This chapter is concerned with the principles involved in the construction of these packages or containers which endow viruses with the unique property of maintaining their genetic competence for prolonged periods while outside a cell.

Crick & Watson (1956) suggested that the containers of small viruses would probably be built from identical protein sub-units packed together in a regular manner. This suggestion was based on the observation that the size of the virus genome is not sufficient to contain information for the synthesis of the large amount of coat protein required if the individual polypeptides were all different. It was assumed that only the viral nucleic acid would code for virus protein; these early predictions have been firmly established for a large variety of simple viruses.

Small viruses like TMV certainly do not contain sufficient information to code for over 2000 different polypeptides and the observable structures are explained by assuming that all the protein units are identical. The construction of containers from identical units presents an interesting architectural problem and has important applications not only to virus structures but also to the production of other cellular components.

At first sight, it might be thought that the containers could be produced in numerous ways, particularly because of the great variety of viral morphology. On the contrary, only a limited number of designs can be constructed from a large number of sub-units. There appear to be two main limiting factors: one concerns the tertiary structure of a coat protein sub-unit as determined by its amino-acid sequence and the other is the quaternary structure resulting from self-assembly of the subunits.

5.2. The influence of self-assembly on virus design

The evidence for self-assembly in viruses will be discussed more fully in §5.7 but it is useful to illustrate here the restriction that such a process imposes on a system. If, for example, a bricklayer only used one method of laying bricks and did not introduce the flexibility required for forming corners, windows and doors, the types of buildings he could produce would be very limited. In fact, the workman would be allowing the nature or shape or bricks to dictate where each one was placed, rather than making a decision himself. This is exactly what happens in self-assembly processes; the instructions are built into the individual sub-units and hence novelty and flexibility are excluded. Consequently, only two main structural designs are found in viruses and the same shapes exist in unrelated groups of viruses. Thus, classifications that are based on morphology must be regarded with caution; although closely related viruses are likely to have similar structures, the converse is not true. As already mentioned (§2.1) viruses are usually rod-shaped or spheroidal. The rod-shaped particles may be rigid or flexible and have helical symmetry. The spheroidal viruses have icosahedral symmetry.

Although self-assembly limits the variety of possible structures, the process possesses two attributes of particular importance to virus and other biological systems, economy and efficiency. By analogy, pressures of cost and efficiency in the modern building construction industry are overcome in part by use of pre-formed units which can be rapidly and cheaply assembled at the building site. Two processes are involved, namely, the construction of the basic units in the factory followed by their rapid assembly into complex buildings.

Two separate processes also operate in biological systems. Unit construction in the factory is analogous to the synthesis of individual protein units from amino acids. This depends on the instructions contained in the base sequence of the nucleic acids. The second process is independent of exterior instructions, since the information necessary for constructing the complexes is built into the individual components. This mechanism has the advantage that control can be achieved at each level of organisation; any incorrectly constructed sub-units are rejected automatically in the assembly process. Thus, complex systems can be built up efficiently and accurately.

In many ways, self-assembly resembles crystallisation. A virus is not a crystal, however, since it contains at least two chemically distinct components. The protein sub-units and the nucleic acid strand(s) come together to form a virus particle presumably because the latter has a lower energy state than those of the components. The process depends on the formation of bonds between sub-units and, as in normal crystal formation, the regularity of the final structure is a necessary consequence of the thermodynamic compulsion to form the maximum number of non-covalent links between the individual units. Whereas in a crystal all the molecules are in identical environments, this restriction need not apply rigorously to structures made up of protein sub-units.

The folding of a polypeptide chain into a tertiary structure depends on the presence of specific amino-acid side chains and, in particular, on the tendency of apolar groups to cluster by **hydrophobic** bonding. When a protein is completely surrounded by an aqueous environment, there is a tendency for most of the apolar groups to cluster in the interior of the protein molecule. Likewise, hydrophobic areas at the surface of protein molecules tend to come into juxtaposition and account for some protein–protein interactions. Water molecules are highly organised in a low entropy state around the apolar regions and there is a thermodynamic driving force to expel water molecules by the clustering of the hydrophobic regions of similar molecules. Our present knowledge of virus proteins is not yet sufficiently extensive to allow a dogmatic statement to be made on this point, but the amino-acid sequence and conformation of the protein of TMV is

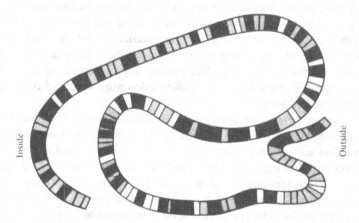

Fig. 5.1. Amino acid sequence of the capsid protein of TMV. The polypeptide contains 158 amino acids. Black areas, amino acids with hydrophobic side chains; stippled areas, amino acids with uncharged polar groups, white areas, amino acids with charged polar groups. (Modified from Fraenkel-Conrat, H. (1969). *The chemistry and biology of viruses*, Academic Press, New York & London.)

known (fig. 5.1) and illustrates this point in convincing manner. It can be seen that at least half of the amino acids present in the inner portion of the TMV protein are apolar whereas on the surface there are only four apolar side chains in a segment of 24 residues. Hence, the same basic physicochemical principles which govern the development of the tertiary and quaternary structures of proteins are responsible for the organisation of virus capsids. The high degree of quaternary structure in virus capsids explains their stability and resistance to various agents and enzymes. The multiple bonds formed during the aggregation of protein sub-units contributes to the masking of enzyme-susceptible sites and to stabilisation against heat and other agents.

The attainment of maximum stability is therefore the overriding factor which determines the final structure of a virus particle; in 1962 Casper & Klug realised that this may involve the introduction of non-equivalent bonding arrangements between identical sub-units, in contrast to a crystal lattice in which all identical molecules are in exactly equivalent environments (fig. 5.2). Casper & Klug described this situation by the term 'quasi-equivalence' which may be defined as 'any small non-random variation in a regular bonding pattern which leads to a more stable structure than strictly equivalent bonding'. The ability of identical proteins to undergo self-assembly in a quasi-equivalent manner makes it possible to build up large virus structures from many relatively small identical sub-units.

(a)   (b)

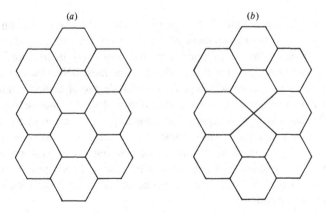

Fig. 5.2. Alternative bonding arrangements. (a) Equivalent bonding in which all the internal bonds are equal; (b) Quasi-equivalent bonding allows the maintenance of the overall design but some of the internal bonds are different.

5.3. Reasons for helical and icosahedral symmetry

It is logical to ask how much influence the nucleic acid has on the nature of the container. Table 5.1 shows that only some single-stranded RNA viruses possess helical structure whereas the other RNA viruses and prac-

TABLE 5.1. *Symmetry of nucleocapsid in different virus groups*

Virus group	Type of nucleic acid	Symmetry of nucleocapsid
Parvoviruses	DNA	Icosahedral
Papovavirus	DNA	Icosahedral
Adenovirus	DNA	Icosahedral
Herpes virus	DNA	Icosahedral
Pox virus	DNA	Complex
Picornavirus	RNA	Icosahedral
Reovirus	RNA	Icosahedral
Togoviruses	RNA	Icosahedral
Orthomyxovirus	RNA	Helical
Paramyxovirus	RNA	Helical
Rhabdovirus	RNA	Helical
Rigidoviridae	RNA	Helical

tically all double-stranded DNA viruses have icosahedral symmetry. A cylindrical container allows more contact between the nucleic acid and the protein sub-units than a spherical container especially if the core is in a compact conformation. It is well established that most DNA viruses contain circular DNA molecules which possess a high degree of coiling and hence probably present a relatively compact globular structure, most easily contained within a spherical capsid. The same argument may apply to single-stranded RNA molecules whose most stable conformation is achieved by a relatively high degree of intramolecular bonding. In the case of helical viruses, direct contact between the nucleic acid and each protein unit is achieved and possibilities of intramolecular bonding within the nucleic acid itself are eliminated.

5.4. Helical symmetry (TMV)

The simplest protective structure than can be made from a large number of identical sub-units is a cylinder composed of a series of rings, each made from a precise number of units. The interaction of singlets or discs, however, could take place in a number of ways introducing different bonding arrangements as shown in fig. 5.3. The advantages for the assembly of a helical construction are obvious, since every single unit is in an identical position.

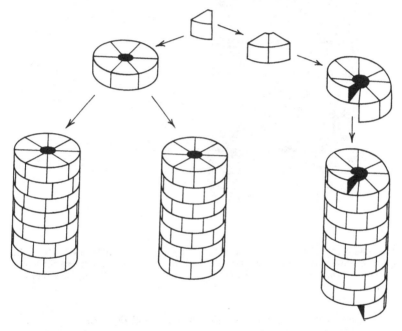

Fig. 5.3. Alternative methods of building a cylinder from identical units.

Tobacco mosaic virus is the most thoroughly documented example of a rod-like virus and most of our current ideas about helical symmetry have been based on the **X-ray diffraction analysis**. A TMV particle is 300 nm in length and 18 nm in width. The single-stranded nucleic acid has a molecular weight of 2×10^6. There are 2200 protein sub-units, each having a molecular weight of approximately 17 000, and these are packed in a helical rod in such a way that each turn contains 16⅓ sub-units and the same disposition of the sub-units is reproduced every three turns. Fig. 5.4 illustrates how the

RNA strand is threaded through the sub-units and is not merely contained in the central hole of the virus.

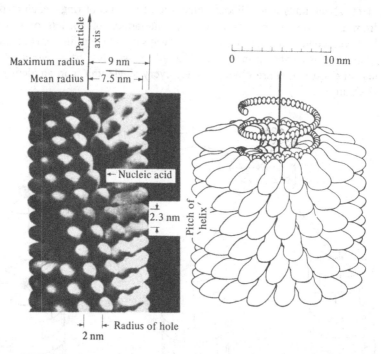

Fig. 5.4. A segment of tobacco mosaic virus particle, model and drawing. Note the external protein molecules and internal RNA helix. (Modified from Fraenkel-Conrat, H. (1969). *The chemistry and biology of viruses*, Academic Press, New York & London.)

Klug and his colleagues have shown that small discs of sub-units are formed initially and they then come together to form a complete helix. The process is illustrated in fig. 5.5 which shows how the pH and ionic strength of the medium affect the way in which the sub-units polymerise. Primary discs are formed by the aggregation of six pentamers which contain layers of two and three sub-units. The pentamers fit together to form a ring known as a double disc which sediments at 20 S. At the appropriate pH and ionic strength the double rings undergo a spontaneous 'shift' so that a helical lattice is produced. Below pH 6, the helical discs polymerise into the cylindrical structure of the mature virus in the presence of RNA. The final assembly is initiated by the interaction of a disc with a special sequence of about 50 nucleotides at the 5'-end of the TMV RNA.

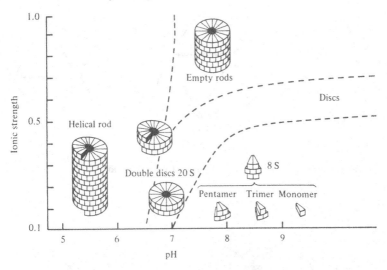

Fig. 5.5. Stepwise maturation of TMV. Diagram showing the ranges over which particular polymeric species of TMV protein participate in maturation. (Modified from Durham, A. C. N., Finch, J. T. & Klug, A. (1971). *Nature, London*, **229**, 37–50.)

The length of a helical virus depends on the length of the nucleic acid, which follows exactly the helical arrangement of the protein sub-units. The overall shape of a helical virus may vary considerably, since the particle may be either rigid or flexible. In the former case, the protein sub-units must exist in exactly equivalent environments; in the latter case, however, flexible or filamentous particles must involve some quasi-equivalent bonding.

5.5. Icosahedral symmetry

Crick & Watson realised that there are only a few ways of building a spherical shell from identical shapes or units. They suggested that a doubting Thomas could convince himself of this 'by trying to draw identical shapes which completely cover the surface of a tennis ball'. The problem is topological and related directly to the optimum design of a shell made up of sub-units. The first evidence for icosahedral symmetry (fig. 5.6) in viruses was provided by X-ray diffraction patterns, obtained by Casper in 1956, of tomato bushy stunt virus and his observations were soon confirmed by Klug, Finch & Franklin in 1957 for TYMV. Fig. 5.7 shows how the X-ray diffraction pattern of poliovirus is similar to that obtained by optical diffraction of an icosahedron containing 60 identical units. The **5:3:2 symmetry** common to

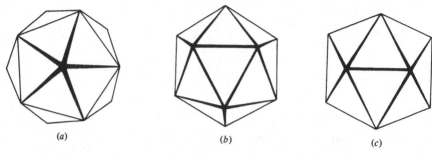

(a)   *(b)*   *(c)*

Fig. 5.6. Icosahedral symmetry. (*a*) 5-fold; (*b*) 3-fold; (*c*) 2-fold.

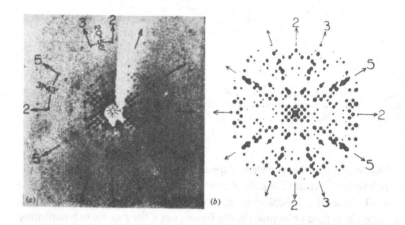

Fig. 5.7. (*a*) An X-ray diffraction pattern of a poliovirus crystal. There are spikes of high intensity along certain directions which are related as the 5- and 3- and 2-fold axes of an icosahedron (as indicated by the arrows). (*b*) An optical diffraction pattern of 60 points on the surface of a sphere with icosahedral symmetry. The intensity distribution of the poliovirus pattern shows the same symmetry relations as this optical analogue. (From Caspar D. L. D. & Klug, A. (1962). *Cold Spring Harbor Symposia on Quantitative Biology*, **27**, 9.)

such particles is easily discernible. Electron microscopy has also played an important part in establishing that icosahedral symmetry is a common feature of spherical viruses. The most conclusive demonstration was made with tipula iridescent virus by use of the technique of 'double shadowing' as shown in fig. 5.8. The shape of the resulting shadows, with one blunt and one pointed, could occur only if the virus particle was icosahedral.

It is now firmly established that the majority of isometric viruses possess icosahedral symmetry and its widespread occurrence suggests that it is not

Fig. 5.8. Tipula iridescent virus and model icosahedron. The virus preparation, (a) was freeze-dried on the electron microscope grid and metallic shadowed from one direction. The grid was then rotated slightly and shadowed again. The two shadows that resulted are identical to the two shadows in (b) which were obtained with an icosahedron photographed with light from two angles. (From Fraenkel-Conrat, H. (1969). *The chemistry and biology of viruses,* Academic Press, New York & London.)

merely a chance, but reflects some underlying structural principle. The reason for a preference for icosahedral symmetry is best seen by trying to construct the various polyhedral structures from identical units such as table-tennis balls. To construct an icosahedron requires 60 identical balls (fig. 5.9). It is impossible, however, to make larger or more complex polyhedral structures in such a way that all the identical sub-units are in equivalent positions. Larger particles can be made in two ways; either by increasing the size of the basic sub-unit, or by increasing the number used. Since the size of proteins are generally in the range 30000–70000 daltons, it becomes a biological necessity to adopt a structure which will be able to accommodate

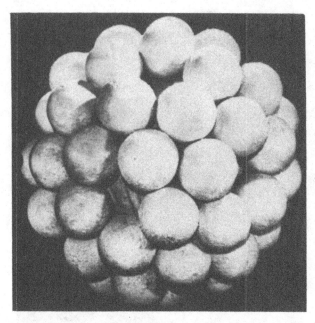

Fig. 5.9. Model of an icosahedron made from 60 balls. (From Stanley, W. M. &
Evans, E. G. (1962). *Viruses and the nature of life*, Methuen & Co. Ltd., London.)

the largest number of structural units and yet allow them to exist in equivalent
or near-equivalent positions. This situation is best achieved with the icosahedron which permits up to 60 asymmetric units to be incorporated into
a spherical framework where they can all have identical environments. Also,
the fact that the size of an icosahedron can be increased by extending the
area of each unit face, thus accommodating more structural units, is a second
major advantage. For obvious geometrical reasons only certain numerical
increases are permitted, as shown in fig. 5.10. Any icosahedron can therefore
be characterised by the number of sub-triangles (T) composing a single face.
There are a number of different ways in which a face of an icosahedron may
be expanded. The simplest situation (known as class $p = 1$) is shown in fig.
5.10(a). Here the primary face can be extended to yield a series of larger
triangles which have **triangulation** (T) numbers equal to 1, 4, 9, 16, 25....
(where T = number of sub-triangles in each face). An alternative primary
face can be a complex of three triangles (known as class $p = 3$) as shown
in fig. 5.10(b). An expansion of this series provides T values of 3, 12, 27...
The T number can be calculated from the equation $T = pf^2$, where f is an

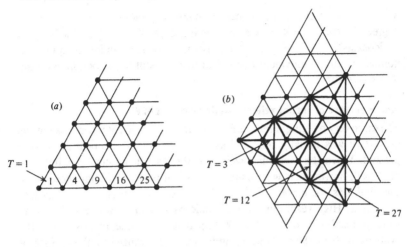

Fig. 5.10. Triangulation numbers. The surface faces of an icosahedron can be built up in a number of ways, although the possible ways are restricted for geometrical reasons. (a) When the primary face is one triangle ($T = 1$) the expanded faces are restricted to 4, 9, 16, 25...sub-triangles. (b) When the primary face has three triangles ($T = 3$) the expanded faces must follow a series containing 12, 27... sub-triangles. There are other more complex arrangements which will not be discussed here.

integer 1, 2, 3, 4, 5 ... and p is the class number. When $T = 1$, the total number of units in an icosahedron is 60 since there are 20 triangular faces, and as seen in fig. 5.9 each unit has an identical position. In structures which are derived from extended unit faces, the total number of units is $60T$. When T is greater than 1, it is impossible to allocate units to exactly equivalent positions and an individual component will be surrounded by either five or six neighbours, since the apices have a five-fold symmetry, as shown in fig. 5.11. As mentioned before (§5.2), the thermodynamic driving force for the self-assembly process is the achievement of the lowest energy state for the

Fig. 5.11. A football. Examine a modern football; it is made from hexagons and pentagons.

whole structure and this becomes the overriding factor in determining the degree of quasi-equivalent bonding that is acceptable. Complex shells based on icosahedral symmetry can be constructed which require the minimum number of quasi-equivalent bonds and hence result in the optimum design of a stable shell.

5.6. Structure of turnip yellow mosaic virus (TYMV)

The discovery by Klug, Finch & Franklin in 1957 that TYMV had icosahedral symmetry was initially regarded as evidence that 60 identical units were present. When the molecular weight of the protein sub-unit was found to be 20000, however, it was realised that 60 of these could not account for the total mass of protein in the shell; in fact, there was a discrepancy of three-fold. High resolution electron microscopy also indicated that there were probably 32 rather than 60 capsomeres. These results posed the problem of how 180 (structural) units could be arranged on the surface of a sphere so that they were able to form 32 capsomeres or 'visible' morphological units. Casper & Klug resolved this by their theory of quasi-equivalence and constructed a model of TYMV in which the 180 units were arranged in two sets of clusters. One set contained 12 sub-sets of five proteins and the other comprised 20 sub-sets of six proteins, corresponding

Fig. 5.12. Turnip-yellow mosaic virus. Model of the virus with its 32 capsomeres (20 hexamers and 12 pentamers of the 20000 molecular weight protein chains = structural sub-units); the cross-section shows the location of the RNA strand. (From Fraenkel-Conrat, H. (1969). *The chemistry and biology of viruses.* Academic Press, New York & London.)

to the apical and facial domains of the icosahedron. Their model is reproduced in fig. 5.12 and illustrates how 32 capsomeres arise by the clustering of the structural units into 12 **pentamers** and 20 **hexamers**. This tendency of the structural units to form oligomers containing either five or six monomers is a common feature of all icosahedral viruses.

5.7. Reconstitution of viruses

The most convincing evidence that virus maturation involves a self-assembly process is provided by experiments on reconstitution. The weak bonds that are responsible for the quaternary nature of virus capsids can often be disrupted to release the sub-units. Thus, many viruses can be degraded to their nucleic acid and protein constituents by exposure to urea, guanidine hydrochloride or acidic conditions.

Under favourable conditions, some virus proteins re-aggregate to produce virus-like particles even in the absence of nucleic acid. Re-aggregation was demonstrated with TMV protein by Fraenkel-Conrat and his colleagues in 1955, and was the first indication that the protein coats of viruses are probably produced by a process of self-assembly.

The source of the nucleic acid seems to have little influence on the aggregation, since RNA from TMV or other viruses or even synthetic polynucleotides can be incorporated into virus-like particles. For example, the RNA of the spherical bacteriophage MS2 can form rod-shaped particles with TMV protein. These rods, however, are only about one third of the length of native TMV. The length of a helical virus particle is apparently determined by the size of the RNA strand.

Although the nucleic acid component does not seem to be an absolute essential for the self-aggregation of the virus coat protein, it does appear to make the structure more stable. In general, particles which are reconstructed in the absence of nucleic acid are less stable than the complete virus. Thus, aggregation of TMV protein occurs at or below pH 5, but the 'empty' rods are dissociated at pH 7. In contrast, reconstituted virus rods (that is, re-aggregates of protein and RNA) are stable over the pH range 2–10 like natural TMV. The process of self-assembly of the protein units appears to be an endothermic reaction, since a similar reversible aggregation can be observed by raising and lowering the temperature between 5 and 25 °C.

Until recently, reconstitution with the proteins of isometric viruses had not been as well documented as with helical viruses. Some small plant viruses, like BMV and CCMV, can be dissociated by raising the salt concentration to 1 M; these proteins re-aggregate to virus-like particles when the salt concentration is lowered by dialysis.

The most clear-cut evidence for the self-assembly of protein sub-units during replication of animal viruses is the production of some particles which do not contain nucleic acid. These are often called **empty particles** or top components. Electron microscopy of negatively stained preparations shows that the stain can penetrate the coat of empty particles to give a ring-like appearance (fig. 5.13).

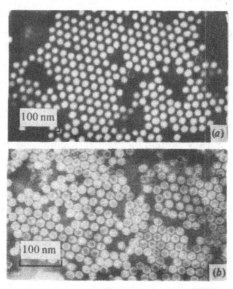

Fig. 5.13. Electron micrograph of bovine enterovirus particles. (*a*) Full particles, i.e., plus RNA, (*b*) Empty particles, i.e., no RNA. (From Johnston, M. D. & Martin, S. J. (1971). *J. gen. Virol.* **11**, 71–9.)

Successful reconstitution of some small bacteriophages has also been achieved and Roberts & Steitz in 1967 obtained infective R17 particles from mixtures of virus capsid proteins and RNA.

It is worth remembering that the reconstitution of a virus is not 'creating life in a test-tube', but it does show that the construction of the virus particle occurs automatically and requires no enzymes or other 'outside' influence. Not all viruses can undergo self-assembly in this manner; the maturation of some of the complex viruses, for example, requires many steps which involve enzymes. With the simpler viruses, however, the construction of the protein capsid appears to be related to the geometric– thermodynamic problem of achieving an optimal design with a minimal energy state.

5.8. Structure–function relationships

The reader may now reasonably ask if a detailed knowledge of virus structure can tell us anything of particular interest to the applied aspects of the subject such as the control of diseases. Clearly, a clinician who is involved with children suffering from poliomyelitis will not need to know whether poliovirus has a shell of icosahedral or helical symmetry, but he does want a suitable and safe vaccine or a potent anti-polio drug. At present there are hundreds of known human viral diseases as well as many economically important diseases of animals and plants. For many diseases, either vaccines have not been produced or are potentially dangerous. It is in this area that research into the detailed chemical structure of the virus capsids will, hopefully, have an applied spin-off. In the first instance the recognition and purification of specific immunising antigens should lead to the elucidation of the precise chemical nature of the antigenic components. Hence virus architecture offers both the intellectual challenge of studying extremely complex systems and the stimulus of working on a socially important topic.

A few essential questions are as follows. How many polypeptide chains are present in the purified virus? Are they all the same or different? How are they located and positioned in the capsid? Which parts of the polypeptides are on the 'inside' and which on the 'outside' of the capsid? Which components have biological activity, for example, antigenic activity or cell receptor activity? In other words, the essence of current work on virus structure is to investigate structure–function relationships.

5.9. Methods of degradation of viruses

The structural sub-units can generally be released from simple viruses by breaking the multiple non-covalent bonds which link the individual units into the symmetrical assembly. Since this inter-unit contact consists mainly of ionic, hydrophobic or hydrogen bonds, techniques which rupture these relatively weak bonds generally cause dissociation of virus particles. There are two main technical problems encountered with dissociation into sub-units. The first is to ensure that the proteins are not irreversibly denatured since this will result in biological inactivation; the best criterion that the protein has not suffered denaturation is provided when it proves possible to reconstitute infective particles but, as we have seen, this is not always practical. The second problem results from the fact that the sub-units are all so alike and have a natural high tendency to aggregate in a specific manner under the environmental conditions of the cell. Under the different conditions imposed by dissociation techniques and subsequent treatments, however, there is a high tendency for non-specific inter-unit bonding which often results in the production of insoluble aggregates.

Another important feature of virus protein is the hidden or masked state of most of the reactive –SH groups. When a virus is dissociated, these groups can readily oxidise to form abnormal complexes. Precautions must therefore be taken to ensure that random re-formation of disulphide bonds does not take place and this is generally achieved by including dithiothreitol or meraceptoethanol in the dissociating solutions. It should be emphasised that there is no single method for dissociating viruses. A method which may be suitable for one virus may be unsuitable for another and the details of many dissociation conditions have been developed by careful empirical research.

The most generally useful method of isolating the individual capsid proteins of even the most stable viruses makes use of strong protein denaturants and heat, conditions which can completely degrade all the weak bonds which hold the shell proteins together. Addition of 8 M urea at pH 7–8 dissociates many viruses completely and others can be degraded by raising the temperature for short periods to 50–100 °C. This is often done in the presence of **sodium dodecyl sulphate (SDS)** which can bind strongly to the individual proteins and may play a part in preventing their aggregation. It also presents the individual polypeptides in a suitable state for electrophoresis on acrylamide gels containing SDS. As mentioned on p. 59 this is a convenient method for determining the molecular weight of the polypeptide(s).

Solutions of concentrated organic acids, such as 67% acetic acid, have been used to dissociate many stable viruses. This acid medium does not cause any deleterious biological effects on protein but acts by the extensive protonisation of all charged groups and by the disruption of hydrogen bonds. If effective, this is often the preferred method since it avoids the use of agents which cause irreversible denaturation of the protein.

The classical method for the isolation of protein from TMV involved dialysis against an alkaline buffer pH 10.5 at 0–4 °C. At this pH, the single –SH groups of TMV protein are not exposed sufficiently to permit autoxidation, hence aggregation of the sub-units is limited. Undegraded virus can be removed by ultra-centrifugation and the protein is precipitated by addition of ammonium sulphate to one-third saturation.

5.10. Selective release of RNA from viruses

The dissociation of many viruses in alkaline conditions is often complex and is related to the ionic strength of the solution. For example, TYMV is dissociated completely to sub-units at low ionic strength but in 0.5 M salt the protein shell remains intact even though the RNA is released. A similar interesting phenomenon is also observed with animal picornaviruses, which can lose their RNA component without destroying the integrity of the capsid shell. This illustrates well the point mentioned in § 5.3 concerning the capsid protein–nucleic acid interaction. The apparent ease of release of nucleic acid from isometric viruses suggests that this interaction is minimal. On the other hand there is no evidence that helical viruses can form artifical empty rods by direct release of RNA, implying a high degree of protein–nucleic acid interaction.

Another feature of the effect of alkali on isometric RNA viruses is that under rigorously controlled conditions the RNA is released in a high molecular weight form and has not been degraded to nucleotides. However, some cleavage has occurred but this appears to be quite specific as the RNA fractions show a single predominant molecular weight. TYMV RNA appears to be cleaved to units of approximately 57000 daltons which correspond to approximately $^1/_{32}$ the size of the intact RNA. This is particularly interesting since as we have seen that the icosahedron structure of TYMV is composed of 32 domains or 'capsomeres' (that is 20 faces and 12 apices). This type of information might indicate that the RNA strand has an interior symmetrical conformation which follows the icosahedral symmetry of the capsid. If so, the parts which approach the surface would be more susceptible to alkaline degradation or nuclease action than the regions buried deep inside the particle.

5.11. Viruses containing non-identical proteins

So far we have considered the structural aspects of viruses which contain only a single type of protein sub-unit. Many important icosahedral viruses exist however which contain non-identical protein components and this fact alone presents further intriguing problems about virus architecture. We shall consider now in some detail the properties of two important groups of animal viruses, picornaviruses and adenoviruses, which illustrate the high degree of complexity and variety that can be found even within the limits of icosahedral symmetry.

Picornaviruses are among the smallest animal viruses and have a diameter of approximately 25 nm. The family includes well over 200 serologically

58   *Architecture of viruses*

distinct types and includes many viruses of clinical importance such as
poliovirus, common cold virus and foot-and-mouth-disease virus. All members of the group so far studied contain at least four non-identical proteins.
The aggregate molecular weight of these proteins is approximately 100000
while the individual polypeptides have molecular weights in the range
10000–40000.

X-ray diffraction studies (see fig. 5.7) suggest that the particles have
icosahedral symmetry and electron microscopy indicates that there are 32
capsomeres. It was initially assumed that the picornaviruses would be like
TYMV and contain only one type of polypeptide and the early reports that
multiple proteins were present were received with some scepticism.
Identification of the N-terminal amino acids and examination of amino-acid
compositions, tryptic digest finger-printing data and electrophoretic properties have confirmed, however, that the polypeptides are truly non-identical.
Electrophoresis in SDS–acrylamide gels has been valuable in this type of
work, since it permits a determination of the number of polypeptide chains
present in a virus particle. Such an experiment generally involves the growth

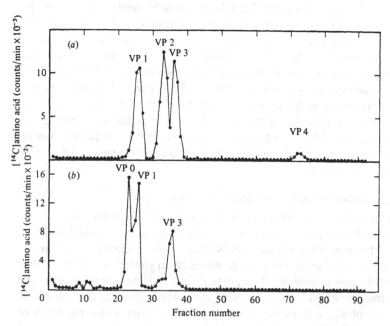

Fig. 5.14. (a) Acrylamide gel pattern of proteins from virus particles (b) Acrylamide
gel pattern of proteins from procapsids. (From Johnston, M. D. & Martin, S. J. (1971).
J. gen. Virol. **11**, 71–9.)

of the virus in a medium containing radioactive amino acids. The virus is highly purified and degraded to individual polypeptide chains by treatment with urea and SDS at high temperatures. Fig. 5.14 shows the profiles obtained after electrophoresis on polyacrylamide gels of proteins isolated from a bovine enterovirus. There are four distinct peaks, VP 1, VP 2, VP 3 and VP 4. The radioactivity associated with each band is proportional to the mass of the protein present and the molar ratios can be calculated by dividing by the molecular weights of the individual proteins. If the molecular weight of the total protein in the virus particle is known, it is possible to calculate the number of polypeptide chains of each type that are present. Thus, it has been established that picornaviruses contain 60 chains of VP 1, VP 2 and VP 3 and either 60 or 30 chains of VP 4. In table 5.2 the results from an experiment on a bovine enterovirus illustrates a method for determining the number of polypeptides present.

TABLE 5.2. *Structural proteins of bovine enterovirus particles and procapsids*

Polypeptide	Virus particles				Procapsids		
	VP1	VP2	VP3	VP4	VP0	VP1	VP3
Molecular weight	34000	28000	26000	9000	37000	34000	26000
Fraction of ^{14}C	0.334	0.322	0.295	0.047	0.384	0.355	0.260
Fraction of stain (i)[a]	0.364	0.306	0.279	0.051	0.381	0.349	0.268
Fraction of stain (ii)[b]	0.370	0.300	0.284	0.046	0.390	0.350	0.260
Average molar ratio	1.05	1.10	1.10	0.53	1.04	1.03	1.01
Probable number of chains/particle	60	60	60	30	60	60	60

After purification, the particles were disrupted by boiling in 1% SDS and 0.8 M urea and separated by electrophoresis on polyacrylamide gels.
[a] Determined by scanning the gels.
[b] Determined by eluting the stain.

5.12. The protomer concept

In 1969, Rueckert and his colleagues made an important observation regarding the structure of the Mouse–Elberfeld (ME) picornavirus. They showed that the virus particle could be dissociated in a stepwise manner by treatment with acetic acid or urea. Two types of protein unit were formed which were studied by sedimentation on sucrose gradients. Their sedimentation values were found to be 14 S and 5 S respectively. The 5 S sub-unit was shown to contain three non-identical chains analogous to VP 1, VP 2 and VP 3 in bovine enterovirus. The molecular weight of the 5 S sub-unit was shown to be 86 000 and 14 S sub-unit was 420 000. This information was consistent with the 14 S component being an oligomer comprising five 5 S sub-units or protomers. Similar calculations to those described in table 5.2 indicated that the intact capsid contains 60 protomers or 180 chains. Rueckert proposed a model for ME virus in which each of the 60 protomers has an identical environment. Therefore, if we consider the protomer as the essential building unit there is no need for quasi-equivalent bonding and the construction is the same as shown in fig. 5.9.

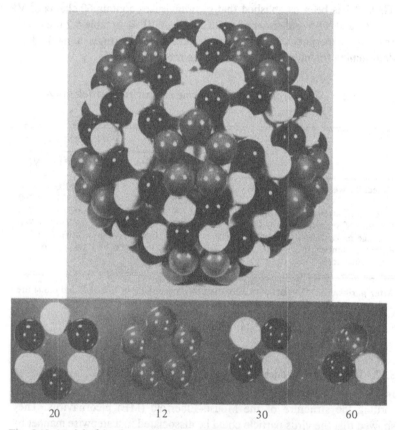

20   12   30   60

Fig. 5.15. Model of bovine enterovirus. 180 balls are arranged in 60 trimers. Each ball represents the location of one of the three structural proteins. The lower part of the photograph shows the number of times the various domains are repeated in the shell. (From Johnston, M. D. & Martin, S. J. (1971). *J. gen. Virol.* **11**, 71–9.)

The existence, however, of three non-identical polypeptide chains in each protomer poses an intriguing question. Are the different polypeptides wrapped up in a complex manner resulting in a single trimeric unit or does each chain have a specific and discrete location, related by specific bonding patterns to its nearest neighbours? Although it is impossible to decide on this issue at present, the weight of evidence favours the latter interpretation. For example, in numerous other multiple protein systems the individual polypeptide chains appear to preserve a high degree of integrity and in picornaviruses, too, there is little evidence that inter-polypeptide –S–S– bonds are present. We constructed a model of a bovine enterovirus from coloured table-tennis balls, each ball representing any one of the three types of protein (VP 1, VP 2 or VP 3) (fig. 5.15). Basically, this gives a structure identical to that of TYMV, except that the individual units are not identical. They are grouped in 32 domains, comprising 12 pentamers (grey balls) and 20 hexamers (black and white balls). In contrast to the TYMV model however, we see here that any particular unit is always in an identical environment. The multiplicity of structural units overcomes the difficulty imposed upon the TYMV system in which identical units must have slightly different bonding arrangements and are therefore quasi-equivalent. In this model of picornaviruses, each identical unit can be equivalent, either as a unit protomer (60) or as individual polypeptide chains (60 of each).

5.13. The locality of specific polypeptides

It is not yet possible to decide unambiguously the specific locality of individual polypeptides, but some circumstantial evidence suggests that at least one of the chains may have a precise locality. Independent experiments in this laboratory on bovine enteroviruses and in Japan with poliovirus have shown that the bonding between VP 2 and VP 1 and VP 3 is susceptible to disruption by treatment with alkali or urea under carefully controlled conditions. The residual particles produced by the release of VP 2 from poliovirus are thought to be a spherical matrix of VP 1 and VP 3 and our own studies of similar particles obtained from bovine enterovirus show ring-like components in the electron microscope. The illustration in fig. 5.16 shows that it is possible to remove completely the apical region (grey balls) and leave a partially stable matrix of white and black balls. The polypeptide VP 2 may therefore be located at the apex of the icosahedron shell, but other possible localities cannot be completely excluded at present.

Another method which can be used to study the locality of proteins makes use of labelling specific surface amino acids. Tyrosine residues, for example, are easily iodinated either by use of an enzyme lactoperoxidase or by

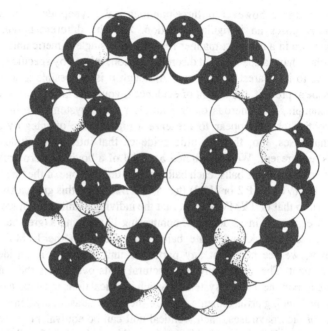

Fig. 5.16. Model of partially degraded particle. The 60 grey balls in Fig. 5.15 have been removed. The basic matrix of white and black balls remain intact. As the protein VP 2 is easily removed it is thought that it is located at the apices.

addition of iodine ion and an oxidising agent such as chloramine T. If radioactive ^{125}I is used, the surface tyrosine residues can be easily identified. Recent studies with FMDV and with bovine enteroviruses (fig. 5.17) show that in iodination experiments only VP 1 is labelled with ^{125}I and hence may have biological surface functions, such as antigenicity or cell receptor activity.

5.14. The assembly of picornaviruses
The model proposed by Rueckert based on 60 identical protomers is of great importance in helping us to understand how these capsids may be assembled. Recent work on the maturation of poliovirus indicates that the immediate precursors of the virus particles are empty shells which contain three polypeptide chains generally called VP 0, VP 1 and VP 3. The empty particles have a sedimentation value of 74 S compared to the mature particle which sediments at 160 S. By some as yet unknown mechanism the viral RNA is inserted into this empty structure and the VP 0 polypeptide is cleaved to form VP 2 and VP 4. Recently, Baltimore *et al.* and Hoey &

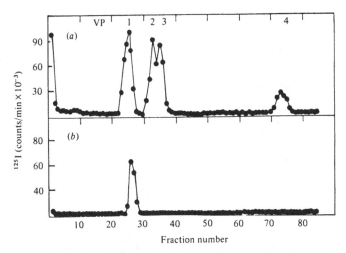

Fig. 5.17. Electrophoresis on polyacrylamide-SDS gels of ^{125}I-labelled virus protein. (*a*) Virus particles were disrupted with SDS+urea at 100 °C for 2 min and then treated with [^{125}I]NaI+chloramine T. (*b*) Virus particles were treated with [^{125}I]NaI+chloramine T and recycled on a sucrose density gradient. The 165 S component was isolated and disrupted with SDS+urea at 100 °C prior to electrophoresis. (From Carthew, P. & Martin, S. J. (1974). *J. gen. Virol.* **24**, 525–34.)

Martin in 1974 have found that other particles exist which have sedimentation values intermediate between the procapsids and the mature virons. These particles also contain RNA and are thought to be provirions in which none or only some of the VP 0 proteins are cleaved. It is likely that the maturation of picornaviruses takes place in a stepwise manner involving a number of stable or partially stable intermediates. Although minor differences may be found in different virus groups the following generalised scheme illustrates the events which probably occur during the final maturation of an enterovirus (fig. 5.18).

Once again we see that the process of self-assembly takes place in a stepwise manner and even when non-identical proteins are present, the structure of the particles can be understood within the general framework of architectural principles outlined earlier. Perhaps it is worth emphasising again that in these multi-protein systems all the units are in identical environments and quasi-equivalent bonding is minimal. Although this may have some advantages, it is extremely expensive from the point of view of the amount of genetic information required. For example, in a picornavirus approximately 35% of the genome is used for the synthesis of structural protein whereas in TYMV, where the sub-unit has a molecular unit of 20000, only 8% of the genome needs to be used for this purpose.

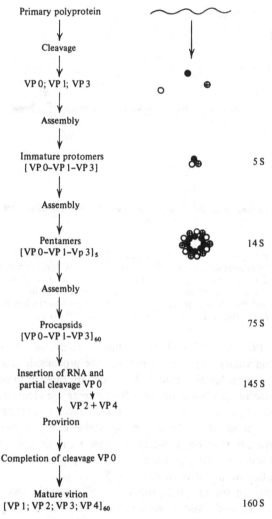

Primary polyprotein

↓

Cleavage

↓

VP 0; VP 1; VP 3

↓

Assembly

↓

Immature protomers   5 S
[VP 0–VP 1–VP 3]

↓

Assembly

↓

Pentamers   14 S
[VP 0–VP 1–Vp 3]$_5$

↓

Assembly

↓

Procapsids   75 S
[VP 0–VP 1–VP 3]$_{60}$

↓

Insertion of RNA and   145 S
partial cleavage VP 0
↓ VP 2 + VP 4

Provirion

↓

Completion of cleavage VP 0

↓

Mature virion   160 S
[VP 1; VP 2; VP 3; VP 4]$_{60}$

Fig. 5.18. Scheme for the maturation of an enterovirus.

5.15. Structure of adenoviruses

In the last section, I have described the methods involved in studying the
inter-relationships of different polypeptides in multi-protein viruses and of
course these general principles have been utilised for many virus types. The
most successful and extensive structural studies on icosahedral viruses have

been done on adenoviruses which I shall use here to illustrate how attempts can be made to relate structural components to biological function. The capsomeres of adenovirus are sufficiently large to be seen clearly in electron micrographs (fig. 5.19). Adenovirus merits some discussion of the calculation of the number of capsomeres that can be present in complex icosahedral shells.

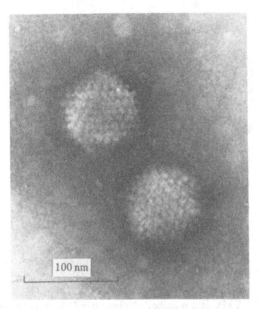

100 nm

Fig. 5.19. An electron micrograph of two adenovirus particles showing capsomeres arranged as an icosahedron and fibres projecting radially from the pentons. (Photograph provided by B. Adair, Veterinary Research Laboratory, Belfast.)

An empirical formula exists which satisfies our requirements and allows the number of capsomeres to be calculated from electron micrographs, provided one face can be clearly seen. The total number of capsomeres is given by $10x(n-1)^2+2$, where n is the number of units on the edge of the virus and x is an integer, 1 or 3, determined by the series class $p = 1$ or $p = 3$ as described on p. 50. The reader should now compare carefully the models of TYMV (fig. 5.12) and adenovirus (fig. 5.20) remembering that in the model of TYMV each individual 'blob' is a structural unit whereas the capsomeres are clusters of either five or six units. In the model of adenovirus each ball is a capsomere. Now, the distribution of the capsomeres can be related to the triangulation grid shown in fig. 5.10. Adenovirus belongs to

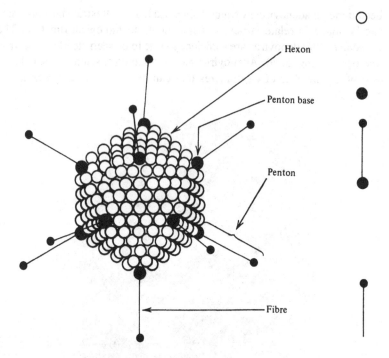

Fig. 5.20. A model of the adenovirus particle based on electron microscopic observations. (From Valentine, R. C. & Pereira, H. G. (1965). *J. mol. Biol.* **13**, 13.)

class $p = 1$ and TYMV is a member of class $p = 3$. It can also be seen that there are two capsomeres present on each edge of a face in TYMV; in contrast, there are six in adenovirus. Table 5.3 shows the number of capsomeres theoretically possible in particles with cubic symmetry. Only the values with bold-faced numbers have been found in viruses. Adenovirus contains 252 capsomeres and TYMV has 32.

Not all the capsomeres in adenoviruses are made of identical structural units and the vertex has an additional fibre component. The 240 capsomeres seen on the faces of the icosahedron are called **hexons** and the 12 units at the vertices are **pentons**. Each penton can be separated into a penton base and a **fibre**. Three types of bonding arrangements are involved: those between a hexon and six other hexons; those between a hexon and six other hexons at an edge between two adjacent triangular faces and those between a hexon and the penton. Adenovirus provides a good example of how

TABLE 5.3. *The possible number of capsomeres in virus particles with icosahedral symmetry*

p = 1			p = 3		
n	C	Virus example	n	C	Virus example
2	12	Phage φX174	2	32	TYMV
3	42	—	3	122	—
4	92	Reovirus	4	272	—
5	162	Herpes virus	5	482	—
6	252	Adenovirus	6	752	—
7	362	—	7	1082	—
8	492	—			
9	642	—			
10	812	—			

C is the number of capsomeres on the virion and is given by the formula $C = 10p(n-1)^2 + 2$.
n is the number of units on an edge of the virion. Bold numbers have been found in viruses.

quasi-equivalent bonding becomes an important feature in the construction of large virus capsids.

Extensive studies have shown that at least five different polypeptide chains are present in the intact virus particle. Two are associated with the DNA in the form of internal components and the remainder are capsid components. Table 5.4 summarises the essential features known at present about the 'fine' structure of adenovirus. Each hexon contains six protein molecules (mol. wt 120 000) and each penton base contains five chains (mol. wt 70 000). The complete virus particle therefore is composed of at least 1500 structural units. The number of structural units in an icosahedron is given by the term $60T$ (see p. 51), where the T number for adenovirus is $1500/60 = 25$. The following derivation shows how the T value for structural units can be

TABLE 5.4. *Capsid components of adenovirus*

Name	Number per virus particle	Mol. wt	Number per morpho-logical unit	Mol. wt	Number per virus particle
Hexon	240	360 000	6	120 000	1 440
Penton	12	280 000	—	—	—
Penton base	12	210 000	5	70 000	60
Penton fibre	12	70 000	1	62 000	12

related to the total number of capsomeres. The total number of units can be given by the sum of 12 pentamers and x hexamers.

$$(12\times5)+(x\times6) = 60T \qquad (5.1)$$

The total number of capsomeres is therefore equal to

$$(x+12) \qquad (5.2)$$

From (i) $60+6x = 60T$

Therefore $x = 10T-10$ (number of hexamers)

Hence $x+12 = 10T+2$

Total number of capsomeres $= 10T+2$
For adenovirus where $T = 25$, the number of capsomeres is therefore 252.

5.16. The biological function of adenovirus components

The biological functions of the various structural components of adenoviruses have been studied extensively. Haemagglutinating activity is associated with penton and fibre and is of particular interest in view of the immunising capacity of preparations of purified pentons. Other capsid components carry antigenic characteristics. The hexons contain a group-specific CF antigen which has been found to be shared not only with human adenoviruses but also with other mammalian adenoviruses although not with the adenoviruses of birds. These immunological cross-reactions are greatest with hexons, less with pentons and highly species-specific with fibre. It has been suggested that the degree of immunological relationship between different kinds of structural components may be related inversely to the degree of their evolutionary specialisation and that differentiation from 'primitive' hexons → pentons → fibres may have occurred. If this is so, studies on the amino-acid sequences of the polypeptides of different structural components should reveal the occurrence of identical regions.

5.17. The structure of complex viruses

Complex viruses do not conform to the typical helical or icosahedral symmetry. In general, these particles contain a membrane which surrounds a nucleocapsid. As mentioned on p. 13 the nucleocapsid components can have helical or icosahedral symmetry. We are concerned here with the composition and structure of the envelope. It has been found that a substantial portion of the membrane is derived from the cellular membranes of

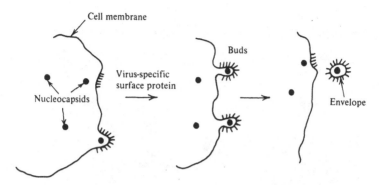

Fig. 5.21. Maturation by budding. The virus nucleocapsid gathers on the inside of the cell membrane and buds out through it, surrounding itself with an envelope. The lipid content of the envelope is derived from the host cell, but the surface proteins are specified by the virus.

the host cell during the final stages of virus maturation. Briefly, as the virus nucleocapsid leaves the cell it becomes surrounded by a piece of cellular membrane by a process known as **budding** (fig. 5.21). Before this happens, the cell membrane is considerably modified at the instruction of the virus, but the basic matrix of the membrane, the lipid and the glycolipid component is of cellular origin. In general, a close similarity exists between the composition of the lipid components of a number of viruses and those of the host cells. The study of virus membranes therefore can tell us quite a lot about the structure of cell membranes in general.

It is impossible in a book of this nature to detail thoroughly all the varied membrane-containing viruses and I shall only describe briefly what is known about the structure of paramyxoviruses. Although it would be incorrect to say that this is typical of all the complex viruses, the methodology and approach to the study of the structure of other groups is essentially the same.

5.18. Structure of paramyxoviruses

This group contains many important viruses such as measles and Newcastle disease. Measles-virus particles are found to be pleomorphic in size, about 100 nm to 300 nm in diameter (fig. 5.22a). The internal component is the nucleocapsid which consists of a single strand of RNA (6×10^6 daltons) complexed with about 2000 molecules of nucleocapsid protein which are about 60000 daltons. The nucleocapsids are long, filamentous, helical structures 18 nm in diameter (fig. 5.22b). The nucleocapsid is wrapped up tightly in an envelope which is composed of lipid and glycoprotein. The envelope

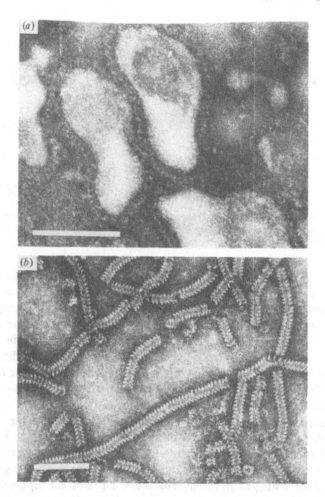

Fig. 5.22. Electron micrograph of (*a*) Measles virus and (*b*) Nucleocapsids. (Photographs supplied by E. Dermott, Department of Microbiology, Belfast.)

is readily broken down by treatment with non-ionic detergents or lipid solvents such as ether.

The glycoproteins are apolar; one end probably extends through the lipid layer to bind in some way with the underlying matrix protein. The hydrophilic ends of the glycoproteins, which probably contain the carbohydrate residues, project from the surface and are referred to as spikes. Treatment of these viruses with proteolytic enzymes often results in the release of the

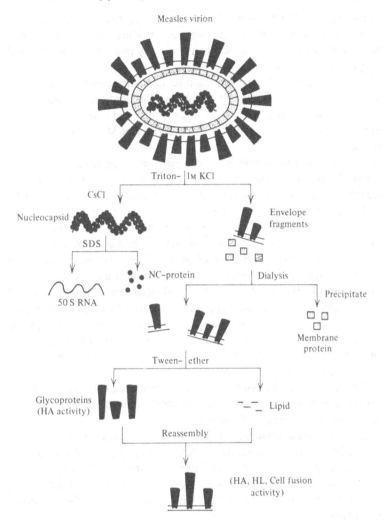

Fig. 5.23. Biochemical dissection of measles virus.

surface glycoproteins to give smooth-surfaced particles which have lost many of the important biological characteristics, such as the ability to haemagglutinate red blood cells or cause haemolysis. In general, paramyxoviruses have two glycoproteins, one of which is the haemagglutinin (HA) and the second is required for haemolytic (HL) and cell-fusion activities. Cell fusion plays an essential role in infectivity and spread of infection from cell to cell as will be described in §6. However, although the glycoproteins are involved in this process, they must be associated with lipid. Active haemolysins can be reconstituted by re-assembly of separated glycoproteins on to lipid.

Separating the lipid layer from the internal nucleocapsid is a sheath of protein known as the matrix or membrane (M) protein. The membrane protein is about 37000 daltons in size, hydrophobic and insoluble in aqueous solutions. There are about 3000–4000 molecules of M protein in a single particle.

The biochemical dissection of enveloped viruses can be carried out in a number of ways but a typical protocol is illustrated in fig. 5.23. The principle used is to disrupt the virion in a solution of a non-ionic detergent and 1 M-KCl which will solubilise the envelope but not degrade the nucleocapsid. The latter can be removed by centrifugation and purified on caesium chloride or sucrose gradients. When the solubilised fraction is dialysed to lower the salt concentration, the membrane protein precipitates. The glycoprotein can then be released from the lipid by extraction with Tween–ether.

A number of other minor proteins are associated with most enveloped viruses. **Polymerases**, for example, are found with the nucleocapsids, and some enveloped viruses contain additional surface proteins, such as the enzyme **neuraminidase**. These minor components do not contribute significantly to the architecture of the particles and will be discussed later in terms of the role they play in replication.

Influenza virus has an essentially similar structure except that the RNA is not a single molecule but is present in several fragments of different sizes. It is said to have a **segmented genome**.

A great deal of our knowledge about enveloped viruses has been obtained from studies on vesicular stomatitis virus (VSV). This is a bullet-shaped particle (see fig. 2.3) which is similar to rabies virus. VSV grows to a high titre in tissue-culture cells and has proved very suitable for biochemical studies particularly in elucidating the chemical nature of the surface antigenic components.

5.19. Structure of poxviruses

The largest and most complex viruses are members of the poxvirus family. The particles are oblong or brick-shaped and are approximately 100 nm × 200 nm × 300 nm. The double-stranded DNA is the largest of any animal virus and measurements of the contour length of electron micrographs of purified DNA show an approximate molecular weight of $150–200 \times 10^6$. The DNA is contained in a nucleoid, shaped like a biconcave disc, surrounded by several

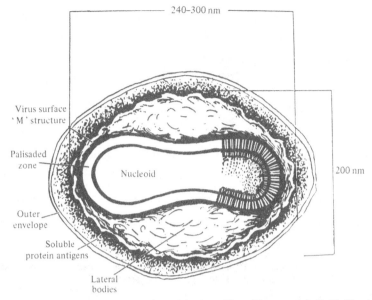

Fig. 5.24. Diagram of the structure of vaccinia virus. (From Westwood, J. C. N., Harris,
W. J., Zurartouw, H. T., Titmuss, D. H. J. & Appleyard, G. (1964). *J. gen Microbiol.*
34, 67.)

layers containing protein and lipid. A diagram of the side elevation of a
particle is shown in fig. 5.24. Complete degradation of poxvirus particles
followed by electrophoresis in acrylamide gels shows that at least 17–20
polypeptides are present.

5.20. Structure of bacteriophages

Although a great variety of morphological shapes exist amongst bacteriophages, they fall into the two basic forms of symmetry, cubic and helical.
Even the most complex structures, like the T-even phages (fig. 5.25), are
composed of two symmetrical components: the heads have cubic symmetry
and the tails have helical symmetry.

Bacteriophages have been classified into five basic morphological types
which are illustrated in fig. 5.26. The most complex particles (the T-even
phages, T 2, T 4, etc.) have contractible tails and a considerable variation
in the shape and size of the heads. The heads have icosahedral symmetry
and have the shape of two halves of an icosahedron connected by a short
hexagonal prism. The T-odd-numbered phages (T 1, T 5, etc.) have long or
short flexible tails with no contractile component.

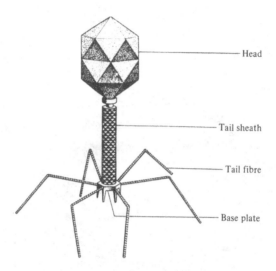

Fig. 5.25. The anatomy of T2 bacteriophage. (From *The genetic code*: The Open University Press, Walton Hall, Bletchley, Buckinghamshire.)

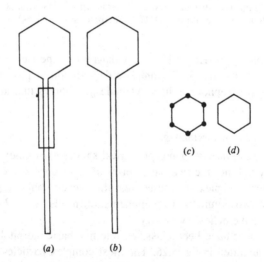

Fig. 5.26. Various morphogenic forms of bacteriophages (*a*) Contractile, (*b*) Non-contractile; (*c* & *d*), icosahedral. (From Bradley, D. E. (1971). *Comparative virology*, Academic Press, New York & London.)

Many bacteriophages do not have tails and are generally small viruses belonging to either of two main groups, depending on the size of the apical capsomeres. ϕX174 is a small bacteriophage containing a single-stranded DNA molecule and at the apices of the icosahedral shell has large capsomeres which are attached to spikes. Some filamentous bacteriophages have also been found in which the DNA is single-stranded.

There is a small group of spherical bacteriophages such as R17 and MS2 containing single-stranded RNA. These are some of the simplest genetic systems known to exist and will be discussed further in the next chapter.

6 The strategy of virus infection

6.1. Introduction

As a great variety of events take place during the process of virus multiplication, it is impossible to describe with any degree of accuracy the details of a typical virus growth cycle and, at the risk of over-simplification, I shall attempt to describe only the salient processes which have universal significance. Naturally, certain viruses have been studied in more detail than others and a general picture of virus replication can only be obtained by the integration of our knowledge of a variety of systems. At the outset it should therefore be realised that the details of what is described for a particular virus may not, and probably do not, apply to other virus groups, although the essential concept involved may be quite general.

The growth cycle of practically all viruses can be divided into a number of discrete steps. For different viruses, the mechanisms used may differ, but all viruses must be able to pass through a fairly precise sequence of events. These discrete events will be discussed in a manner which will build up the general picture of how the latent 'dead' genetic material of the virus nucleic acid can become 'alive' when released into the appropriate genosphere (§1.2). In brief, the virus life cycle involves eight main steps: attachment, penetration, uncoating, control of cellular events, expression of the virus genes, synthesis of virus proteins, replication of the virus nucleic acid, maturation, and the release of progeny virus particles. The transformation of cells by viruses will also be described in §6.5.2.

6.2. Attachment

It is not by coincidence that I have included as a frontispiece a cartoon illustrating the invasion of a planet by the inhabitants of a space ship. For many purposes, the virus coat may be considered as the space vehicle which can convey the virus genetic material safely from one genosphere to another just as future space ships may carry humans from one biosphere to another. Unfortunately, the analogy does not stop here, since, as we shall see, the behaviour of man sometimes mimics only too closely the activities of viruses.

76

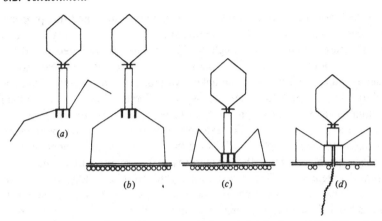

Fig. 6.1. The steps of attachment, penetration, and injection of the T4 and T2
bacteriophages to their host cell. (*a*) An unattached phage. (*b*), the long, thin tail fibres
have attached to the cell wall. They are kinked slightly near their middle. The base
plate of the phage is over 100 nm from the cell wall. (*c*), the phage has moved closer
to the cell wall, and the pins extending from its base plate are in contact with the
wall. Long tail fibres are shown kinked and bent slight at their hinge on the base plate.
(*d*), the tail sheath has contracted, and the sheath has retracted up the tail tube,
carrying with it the freed base plate. The plate is still linked to the host cell by short
tail fibres. The inner tail tube has penetrated the outer three layers of the host cell
wall. A decrease in the density of the rigid portion of the cell wall is also shown. (From
Fraenkel-Conrat (1969). *The chemistry and biology of viruses*. Academic Press, New
York.)

Certain bacteriophages have a highly developed and complex attachment
apparatus, similar in many ways to the shape of a lunar landing module. In
the case of the T-phages, the attachment to the surface of the *E. coli* cells
is effected by the ionic strength of the medium, although they require
specific co-factors such as L-tryptophan. A few molecules of tryptophan bind
to the tail fibre proteins causing their extension in a manner favourable to
their adsorption on to the cell surface. The receptor sites on the cell surface
are highly specific, some being lipoprotein in nature and others
lipopolysaccharides. After attachment of the tail fibres, the hexagonal base
plate moves closer to and becomes attached to the wall (fig. 6.1). At this
point the tail sheath contracts and forces the tail through the soft layers of
cell wall. Final penetration of the inner rigid layer of the cell wall may
require the action of a phage enzyme, **lysozyme**.

Not all viruses attach to cells in such a dramatic and dynamic manner as
is seen with the T-phages, but they all possess a high degree of host
specificity. The small RNA bacteriophages attach on to cellular protrusions
of the bacterial cell known as **F. pili**.

Attachment of viruses is independent of temperature and this property has been used to synchronise infection since the next step (engulfment) is temperature-dependent. Although specific receptors are required in both the host membrane and the virus particles, the attachment appears to be electrostatic as viruses become linked with ionic or van der Waal's bonds of sufficient strength to counteract dispersive forces. When viruses contain a number of different polypeptides in their capsid, attachment appears to involve specific sub-units. For example, it is thought that the pentameric spikes in the bacteriophage ϕX174 are the most likely attachment sites.

The attachment by orthomyxoviruses and paramyxoviruses has been studied extensively. The receptors are glycoproteins. Although some of these viruses contain the enzyme neuraminidase as a surface component, it has been shown that this enzyme is not involved with either attachment or penetration. Influenza virus is not inactivated by antisera against the virus neuraminidase whereas antisera against the haemagglutinin sub-units completely inhibit infection.

Viruses therefore present a wide variety of methods of attachment depending on the presence of precise virus–cell receptor sites. At this early stage, the cell need not respond to or repel the virus. Here we see the first successful step in the strategy of infection: even when a cell is at 0 °C in a quiescent state it is open to attack.

6.3. Penetration

The experiments of Hershey & Chase in 1952 demonstrated that only the nucleic acid of T-phages entered the *E. coli* cell and the virus coat and other appendages were left outside on the surface. In this case, the penetration depends on the ability of the contractile tail of the virus to squeeze the DNA molecule into the cell. This contraction resembles that of muscular action, although the role of ATP in the process has not been clarified. The task is quite enormous since the molecule of DNA is 50000 nm long and 2.4 nm in diameter and has to pass through the tail tube which has an inner diameter of only 2.5–3.0 nm. The penetration of the complete DNA molecule takes about 1 min and the entire growth cycle from infection to lysis of the host cell and release of about 300 progeny particles is over in about 30 min. Recently it has been found that some 'inner' proteins are injected along with the nucleic acid and may play an important role in the subsequent steps although the protein shell remains outside the cell. In contrast, many viruses are taken up by the cell in an intact state, leaving none of the capsid outside. This is probably the case with the small RNA bacteriophages and plant viruses such as TMV. There is now good evidence that most animal viruses are taken into the cell by **phagocytosis**. The natural response of cells to foreign particles is to ingest and dispose of them but, as we shall see, viruses make use of this process to penetrate the cell membrane and they then out-manoeuvre the cellular mechanisms which would otherwise degrade them. Whereas attachment of a virus to a cell is temperature-independent, penetration is a temperature-dependent step and can be inhibited by metabolic poisons like sodium fluoride that block phagocytosis.

The process of penetration by enveloped viruses is very complex. Some enveloped viruses such as paramyxoviruses can cause fusion between similar or different cells. For example, cells containing nuclei from both human and mouse cultures can be formed by exposing a mixed culture to Sendai virus. A common feature of paramyxoviruses such as measles virus is the ability to cause the formation of large multi-nucleated cells known as **syncytia**. This can be achieved by the use of purified envelope components from measles virus, as shown in fig. 6.2.

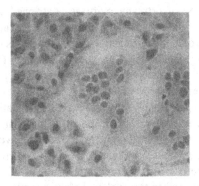

Fig. 6.2. Cell fusion caused by the envelope of measles virus. The large multinucleated cells are called syncytia.

Hence, the classical picture of viral entry into a cell is no longer valid. Only in exceptional cases does the nucleic acid enter alone and the more common pathway is the complete engulfment of the intact virus particle. There are obvious advantages to this since the virus genome may remain protected for some time from the alien genosphere, awaiting appropriate developments or a particular stage in the cell cycle which would favour its replication. By analogy, an encapsulated astronaut will wait and choose his time carefully before setting out to explore a new and alien biosphere.

6.4. Uncoating of viruses

There is a marked delay after intact virus particles are taken up by cells before the virus begins to have an effect. This effect has been demonstrated with TMV infection, when several hours longer are required to obtain the first symptoms than when naked RNA is used as an inoculum. The period during which nothing obvious appears to happen is called the **eclipse** and is of very variable length, ranging from 2 h for picornaviruses to 10–12 h for vaccinia virus.

With animal picornaviruses, the exact mechanism of the release of the viral nucleic acid is still not known. It may take place very rapidly immediately after the virus particles are engulfed. The ease with which the nucleic acid can be removed in the laboratory without the collapse of the capsid structure (see p. 57) suggests that a similar process may occur in the cell. The rate of the uncoating processes appears to be the limiting factor with most other animal viruses and determines the length of the eclipse phase. The envelope of orthomyxoviruses is stripped off while in the β-phagocytic vesicle, but the fate of the liberated nucleocapsid is not known. Where there are two distinct protective protein coats, as in reoviruses, only the outer one is degraded in the β-phagocytic vesicles. The release of viral nucleic acids from the nucleocapsids raises unsolved problems since the ribonucleases present in the vacuoles should rapidly destroy them. At this stage, the virus can presumably inhibit the cell's degradative processes, and it is likely that the nucleocapsids escape from the phagocytic vesicle and somehow appear in the 'replication factories' which have specific localities in either the cytoplasm or nuclei. Some DNA viruses, such as papovaviruses, adenoviruses and herpes virus are replicated in nuclei, and the nucleocapsids are probably transported to the nucleus before completion of uncoating.

The process of uncoating of members of the poxvirus group is extremely complicated but has, in fact, been studied in more detail than has the process in many other viruses. Virus is taken up by phagocytosis and complete viral particles with all their coats intact can be seen in the vacuoles. Uncoating appears to be a two-stage process. The first stage starts immediately after penetration and results in the breakdown of about half the outer envelope material although the central core of DNA is not exposed. It is probably a **host-mediated** process, the β-phagocytic vesicle responding as it would to any foreign particle. However, at this stage, digestion of the virus particles by the host cell ceases and dramatic alterations in the normal behaviour of the vesicles take place. The walls of the vesicles disappear and the virus core lies free in the cytoplasm. The second stage of uncoating involves the release of DNA from the core and is thought to be effected by an uncoating protein which is produced by the virus genome while it is still inside the core. Vaccinia virus particles contain a DNA-dependent RNA polymerase which is responsible for transcribing virus DNA in the core structure and after its release. It seems that the initial stages of uncoating may be under the control of the cell whereas the final stages are controlled by the virus.

Little is known about the immediate fate of the virus nucleic acid once it is released from the nucleocapsid. There is evidence that upon release of RNA molecules from riboviruses the nucleic acid combines with cellular proteins known as **informofers** to produce virus-induced informosomes. In this way, the virus RNA can immediately make use of the cellular apparatus and be protected from nuclease attack probably in the same manner as cellular RNA molecules. Although these virus-induced informosomes can be isolated from cells, it is uncertain whether or not they are involved in replication or are merely artefacts produced during isolation. However, the information available at present suggests that virus genomes are immediately incorporated into the general pattern of processing cellular nucleic acids upon release from the capsid, although the details of the subsequent steps will depend on the function of the new proteins.

There appear to be only a limited number of pathways open to virus growth and we shall look at these aspects of the strategy of infection in the following section.

6.5. The different types of virus infection

It is perhaps useful to pursue still further our analogy of viruses and the behaviour of humans. History tells us that an invading army can deal with a subjected or conquered country in three main ways. In the most primitive type of conquest, the invading forces can rape the country of its wealth and resources, murder, butcher and eliminate the native population and eventually leave the country devastated and ruined. The wandering warlike tribe will then seek out another likely prey and repeat the process. Alternatively, the invading army may rapidly come to some peaceful co-existence with the conquered forces and effectively redirect the interests and policies of the country. Often there can be an apparently successful assimilation of the invading power, although the old regime will never be the same again! A third type of invasion can establish a new stable state by force but one which can only be maintained by the continued presence of armed forces in excessive numbers. In practice, these three sociological states are mutually exclusive.

Three types of invasion or attack can also be found in the behaviour of viruses and furthermore they also appear to be mutually exclusive; thus, two types of infection do not occur simultaneously in a single cell. We shall call these three types of infection, lytic infection, transformation and persistent infection.

6.5.1. Lytic infection. This is the classical type of virus infection and most virus agents have the ability to attack cells in this manner. In general, a virus particle will enter a cell and multiply fairly rapidly so that a large number of progeny particles is formed which can be released from the cells either slowly or rapidly. An *E. coli* cell infected with a T-even phage will produce some 300 new phage particles within about 30 min. We have seen that the

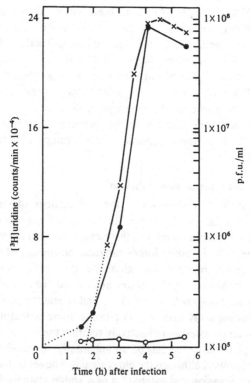

Fig. 6.3. Growth of bovine enterovirus in baby hamster kidney cell cultures. ×——×, infectivity (p.f.u./ml); ●——●, radioactivity in virus-induced RNA; ○——○, radioactivity in RNA extracted from uninfected cells. (From Clements, J. B. & Martin, S. J. (1971). *J. gen. Virol.* **12**, 221–32.)

process is called the **growth cycle**, which is simply defined as the period between the time of infection and the time when new virus is produced. With animal viruses the growth cycle varies considerably in duration. Some picornaviruses, such as FMDV and bovine enteroviruses (fig. 6.3), can produce new virus particles 2 h after infection whereas with other viruses several days are required.

6.5.2. Transformation. In 1911, Peyton Rous discovered a virus which caused tumours in chickens. Since then, many viruses have been found which can transform cells either in animals or in cells in tissue culture. In many cases the infecting particles disappear, although occasionally they can be induced to re-appear and produce lysis of the cells. Certain bacteriophages, known as '**temperate**' phages, can also behave in this manner and remain in a non-obtrusive state for many cell generations. Sometimes, however, these temperate phages can become '**intemperate**' and suddenly begin to multiply and produce progeny particles resulting in the lysis of the bacterial cells. It is now well established that the phage nucleic acid can become **integrated** into the bacterial chromosome and we say that the virus DNA is in the '**prophage**' state. This is discussed in greater detail on p. 108.

The situation is more complex with animal viruses and a variety of virus groups are capable of causing transformation. These are now generally termed **oncogenic viruses** as in some cases they have been shown to cause tumours. Both DNA and RNA viruses, mainly members of the papovaviridae, adenoviridae and retroviridae (leukovirus) families, can cause transformation.

In all cases, it appears that viral genetic information is integrated into the chromosomes of the host cell and few of the virus characteristics are obvious. However, the nature of the transformed cell is drastically altered. More will be said later about oncogenic viruses.

6.5.3. Persistent infections. The third mode of infection is called **persistent** or chronic infection. It is particularly important in animal cells infected with herpes viruses or paramyxoviruses. In this type of infection, the pathway to complete lytic infection appears to be abortive and partially formed virus particles or nucleocapsids accumulate in the cells. In contrast to oncogenic viruses, the cell membranes of persistently infected cells have properties very similar to the envelopes of the virus and, for example, possess haemagglutinating activity and are recognised by virus-specific antibodies. Fig. 6.4 shows how antibodies which are tagged with a fluorescent dye absorb specifically on to cells which are persistently infected with measles virus

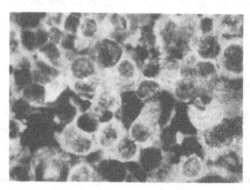

Fig. 6.4. Detection by immunofluorescence of measles antigen in persistently infected cells. (Compare fig. 3.4.)

but do not bind to non-infected cells. The nature of these cells has been drastically altered and not merely by a degree of peaceful co-existence but with an ever-present and obvious persistence of the invading particles.

These three types of infection appear to be mutually exclusive; for example, HEP-2 cells which are persistently infected with measles virus are not susceptible to infection by a strain of measles virus which normally produces a lytic infection. There is a growing body of evidence that suggests that persistently infected cells may play an important role in certain degenerative diseases, especially of the central nervous system.

Before exploring further the details of the expression of the viral genome and the replication mechanisms, it is perhaps of interest to try to relate the three types of infection in a generalised scheme. We have seen that all viruses must attack, penetrate and release the virus genome in a biologically active state and it should be obvious by now that, after the production of new virus nucleic acid and protein molecules, maturation of particles takes place, followed by the release of the progeny virus particles. These individual events all apply to a lytic growth cycle and may be summarised by a series of discrete steps:

$$A \rightarrow B \rightarrow C \rightarrow D \ldots N \rightarrow M \rightarrow R$$

Within this general framework we can see how, at certain stages, a 'switch off', redirection or diversion of the lytic pathway may be effected. In the case of transformation, it would occur fairly early in the growth cycle, for example,

$$A \rightarrow B \rightarrow C:$$

whereas in the case of persistently infected cells the switch off may be relatively late since partially matured virus components are found. In these cells the theoretical pathway may be expressed as

$$A \rightarrow B \rightarrow C \rightarrow D \ldots N:$$

where an important step in the maturation may be inhibited.

6.6. A biochemical coup d'état

The events immediately following the uncoating process depend on the nature of the nucleic acid. In the case of deoxyviruses, the virus DNA must first be transcribed into virus-specific messenger RNA (mRNA) and this then can make use of the cellular ribosomes to effect the synthesis of virus-specific proteins. With riboviruses, however, it is not always necessary for RNA synthesis to occur before virus proteins are produced. For example, picornavirus nucleic acid can act as its own mRNA and immediately after release from the virus can become attached to the cell ribosomes. In some riboviruses, the mRNA is not the virus nucleic acid itself but is the complementary strand. In these instances, for example paramyxoviruses, early synthesis of nucleic acid is essential. The possible schemes are summarised in fig. 6.5 and each will be discussed in greater detail in the following sections.

The central feature of the biosynthesis of virus proteins became clear in the early 1960s with the discovery of mRNA. Viruses do not contain sufficient genetic information to permit the synthesis of ribosomes, transfer RNA (tRNA) molecules and the numerous components essential for protein synthesis. The key factor is that virus-induced mRNA molecules can re-programme the ribosomes and therefore redirect synthesis towards virus 'products' instead of cell 'products'. The first indication of this take-over was provided by experiments with T2 phage and *E. coli* cells. Fig. 6.6 illustrates the general outline of the experiment in which *E. coli* cells were grown in a medium containing ^{15}N (heavy nitrogen). Ribosomes isolated from these cells were more dense than those from non-treated cells. Heavily labelled cells were transferred to 'light' ^{14}N medium so that any ribosomes formed after the transfer could be easily separated from the pretransfer ribosomes. At the same time, the cells were infected with T2 phage and radioactive precursors of proteins and nucleic acids were added. Two important observations were made in this experiment. First, all the newly formed (that is, radioactive) RNA was found associated with the heavy ribosomes. Secondly, the size and composition of the RNA was different from ribosomal RNA (rRNA). These results mean that the infected cell does not make ribosomes after infection and that the virus RNA attaches to and

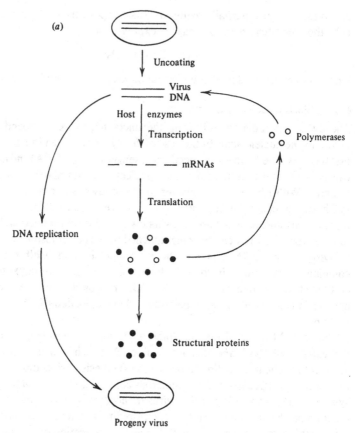

Fig. 6.5. Outline of virus replication and translation. (*a*) Typical DNA virus. (*b*) Variations in the schemes adopted by different RNA viruses. Symbols: v, virion RNA; c, complementary strand of virus RNA; (+) defined as strand which can be used as mRNA; (−) defined as the strand which is complementary to the sequence of bases present in mRNA. Under this scheme, RNA viruses can be classified as **Positive** strand viruses or **Negative** strand viruses.

makes use of the ribosomes that are already present in the cell. These observations had a tremendous impact on our understanding of protein biosynthesis, since it is from here that we can trace the roots of the concept of mRNA. The ability of the invading virus to reprogramme the host cell's ribosomes and redirect and control protein synthesis results in what is essentially a biochemical *coup d'état*. At this point, the strategy of infection changes from a relatively passive role, as seen in attachment,

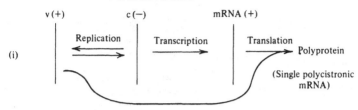

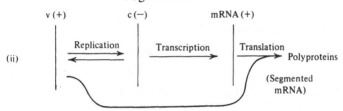

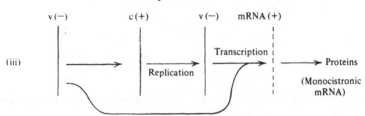

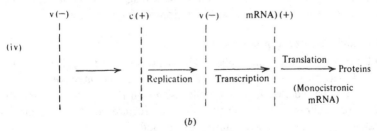

(b)

Fig. 6.5(b).

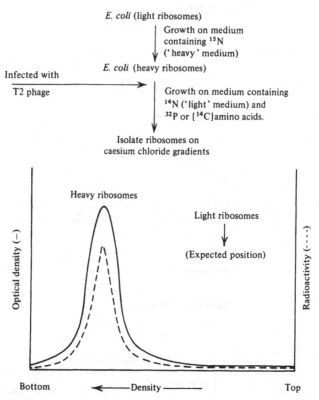

Fig. 6.6. Outline of the type of experiment used to show that viruses reprogramme the ribosomes to produce virus protein.

penetration and uncoating, to an offensive and aggressive position, right at the centre of metabolic control, in the ribosomal system itself.

6.7. Inhibition of cell processes

If the process of infection is followed by measuring infectivity, it is generally found that, once uncoating has been accomplished, no infective virus particles can be isolated from the cells. Although this has been called the 'lag' or 'eclipse' phase, many important features of the biochemical 'take-over' of the cell occur during this period.

One of the most striking events is the inhibition of cellular processes that occasionally takes place soon after infection. For example, within a few minutes after entry of T4 DNA into the bacterium, events are initiated that cause the inhibition of host-cell DNA. This is achieved by the production of an enzyme, deoxycytidine di- and tri-phosphatase which converts one of the immediate precursors of cellular DNA to the monophosphate state (fig. 6.7). The source of this 'new' enzyme will be considered shortly, but we

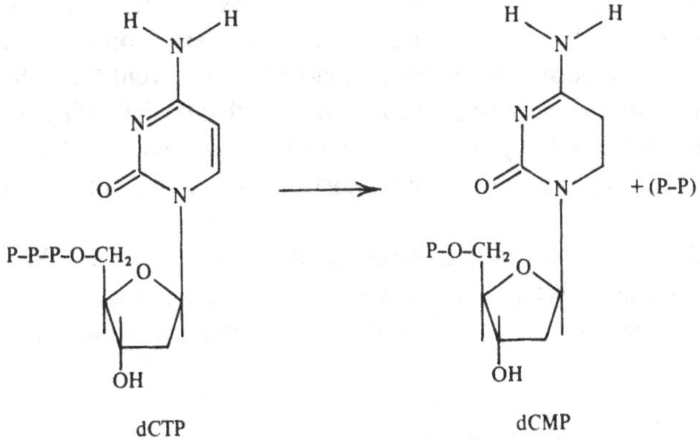

Fig. 6.7. A first step in a biochemical coup d'état.

see here how the phage can strike very efficiently at the centre of cellular metabolism, since by this single step it can switch off the synthesis of the host cell's DNA by removal of an essential precursor. A similar 'switch off' of host-cell DNA, RNA and protein synthesis is found in certain other virus systems. Picornaviruses often inhibit synthesis of macromolecules in the host cell although the chemical details of this process have not been established. The best example is provided by the changes which are rapidly induced by infection of HeLa cells with poliovirus. Two types of inhibition appear to occur. Firstly, there is a decrease in cellular RNA synthesis resulting from the production of a protein which inhibits the host cell's RNA polymerase. Secondly, there is inhibition of cellular protein synthesis by the disintegration of cell polyribosomes which occurs rapidly after infection with poliovirus. Ribosomes, however, can be used for virus-protein synthesis and it is unlikely that the inhibitory action is directly against them. It is more probable that degradation of cellular mRNA molecules occurs.

For any of these inhibitory processes to occur there is a requirement for protein synthesis, and the origin of the inhibitory proteins raises an interesting question. Does the virus genome contain information for the production of 'inhibitor proteins' or does the presence of virus in the cell result in the activation of certain latent control mechanisms which are already present? The virus could be switching on certain negative control processes rather than be switching off synthesis of macromolecules.

Experiments with certain drugs which inhibit poliovirus replication have helped to elucidate this problem. Guanidine is a potent inhibitor of some strains of poliovirus. Although it inhibits the replication of viral RNA, it does not prevent the inhibition of cell-protein synthesis. Hence, guanidine cannot cure cells infected by polio, since death occurs very soon after infection. Another drug, a derivative of thiopyrimidine, can prevent the initiation of host-cell inhibition processes, and it is thought that the inhibiting agents are coded for by the virus genome and probably represent very 'early' virus proteins. The details of most inhibitory processes are not as yet firmly established.

Not all viruses can initiate mechanisms resulting in the inhibition of cellular processes. When a virus causes transformation of a cell, or possibly becomes persistent, it would be of no advantage to inhibit all cellular synthesis.

6.8. Acquisition of new metabolic activity

In 1952, Wyatt & Cohen discovered that the nucleic acid of T-even phages contained an unusual base, **5-hydroxymethyl cytosine (HMC)**. The discovery of HMC led to a great deal of research into how the precursors of phage DNA were formed and what new metabolic steps had to be acquired by infected cells.

The synthesis of HMC depends on a number of enzymes, for example **deoxycytidylate hydroxymethylase**, which are not present in non-infected bacteria. Their appearance poses many challenging questions which are applicable to the general problem of the acquisition of metabolic activity. The origin of the new enzyme activity is the essence of the problem and in theory a number of alternative mechanisms exist for its production. For example, the enzyme may be already present in the cell in an inactive form, such as a zymogen, which is somehow activated by infection. Alternatively, a virus component, or induced product, could cause a derepression of a host-cell gene, in a manner similar to the induction of β-galactosidase. In both these examples the genetic information required for synthesis of the new enzyme is contained in host-cell genes. We have already discussed the possibility that enzymes may be carried into the cell as part of the virus coat. It would be wrong to assume that all 'new' enzymes or proteins induced by infection are translated directly from virus genes. It should be appreciated that the design of experiments to distinguish between these alternatives is both intriguing and complex. For example, the proof that deoxycytidylate

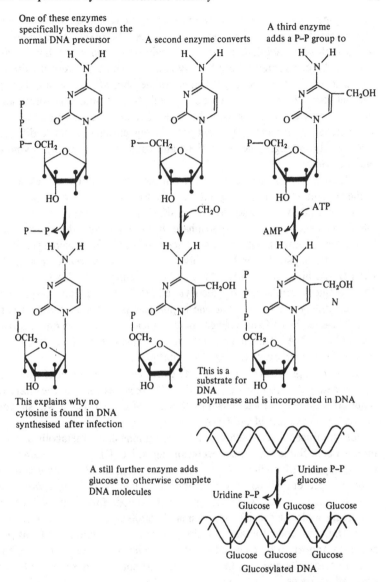

Fig. 6.8. The biochemical mechanism that brings about the synthesis of DNA lacking cytosine and containing instead 5-OH-methylcytosine and its glucose derivatives. Immediately after infection, a number of specific enzymes are synthesised. These are coded for by the viral DNA, and each has a specific role in ensuring the successful multiplication of the virus. (From Watson, J. D. (1970). *The molecular biology of the gene.* W. A. Benjamin, Menlo Park, California.)

hydroxymethylase is coded by the virus genome required over five years' concentrated work by Cohen and his colleagues. Their experiments involved the growth of a methionine-requiring strain of *E. coli* in [¹⁴C]methionine until all the cell proteins were heavily labelled. The cells were then transferred to medium containing [¹²C]methionine and, after ensuring that the internal pools of amino acids were completely free of radioactive precusors, the cultures were infected with T-phage. The enzyme was isolated from 40 l batches of infected cells and extensively purified by three different types of chromatographic separations and found to be virtually devoid of radioactivity.

At least three enzymes are involved in the synthesis of HMC as shown in the scheme illustrated in fig. 6.8. The first enzyme formed breaks down the normal DNA precursor (dCTP to dCMP). This is then converted to 5-hydroxymethylcytosine 5′-phosphate by the enzyme deoxycytidylate hydroxymethylase and other virus-induced enzymes add phosphate residues to form the substrate for DNA polymerase. After the replication of the virus DNA, another enzyme adds glucose to some of the HMC residues. The biological significance of glucosylated DNA in the T4-phage system is not yet clearly established but one hypothesis is that their function is to protect T-even DNA from a phage-specific nuclease which can only degrade unmodified DNA. This hypothesis would explain how *E. coli* DNA is selectively broken down during virus multiplication.

The T-even phages contain about 45 genes and at least eight of these are transcribed within 10 min after infection. These genes code for the enzymes required to alter the metabolism of the *E. coli* cell so that it will support the replication of the phage DNA.

Not all viruses cause such extensive alterations in the metabolism of the infected cell. Some very small bacteriophages, like R17, only contain three genes, capable of coding for a polymerase, an assembly protein and the structural protein. In contrast, the animal virus herpes simplex virus, has a DNA molecule (mol. wt 75×10⁶) which could code for about 100 proteins of average molecular weight approximately 50000. Polyoma virus has sufficient DNA to code for about eight proteins, but a number of enzyme activities appear to be elevated in infected cells compared to the levels in non-infected cells and it is likely that at least some of these are coded by host-cell genomes.

It has previously been mentioned that a virus may produce a product which will alter the function of a host protein. The best example of this is the production of **virus-induced sigma factors** by T-even phages (fig. 6.9). The RNA polymerase present in the infected cell is modified and used by the

(a)

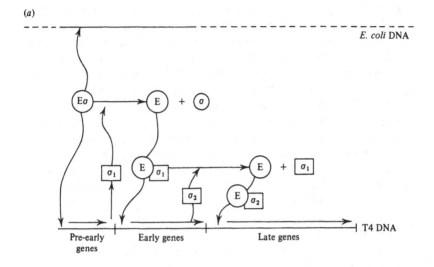

(b)

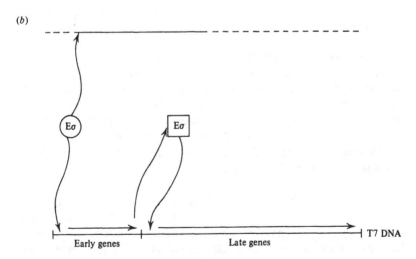

Fig. 6.9. The role of sigma factors in controlling the function of polymerases in infected cells, (a) Cells infected with T4 phage, (b) cells infected with T7-phage. O, host cell; □, virus induced.

virus for the synthesis of virus mRNA. The first gene to be transcribed on the T4 DNA produces a virus-specific sigma factor (1) which replaces the host sigma factor in the polymerase and allows it to transcribe the set of 'early' genes. About 10 min after infection, a second T4 sigma factor (2) is coded for by one of the early genes and this again modifies the RNA polymerase and initiates transcription of the late T4 genes. On the other hand, with T7 phage only the first four genes are controlled by the *E. coli* sigma factor. One of these genes produces a completely new polymerase which transcribes the remaining 25 to 30 genes of the T7 DNA.

A number of other features are particularly relevant to this discussion of the acquisition of new metabolic function. After T4 infection, some changes appear to take place in the host cell's ribosomes so that they function better with virus mRNA than with the host-cell messenger. Also, totally new tRNA molecules have been found in some virus-infected cells and hence viral genes as well as host genes can code for tRNA.

In conclusion, a large variety of events which take place shortly after infection alter particular aspects of the metabolism of the cell. The alterations can result from the direct expression of the virus genome, the modification of the host-cell protein or the repression of host-cell genes. The overall result, however, is the subtle modification of the infected genosphere so that it is more suitable for translating and replicating the invading virus genome.

6.9. Translation of virus genomes

As we have seen, viruses survive in the host cell by being able to reprogramme the ribosomes by addition of virus-specific mRNA. How this mRNA is translated depends on the nature of the host cell. For example, bacterial cells can recognise **internal termination** sequences in polycistronic mRNA whereas eukaryotic cells utilise only **monocistronic** mRNAs.

Four mechanisms are used which were outlined in fig. 6.5 and will be illustrated here by specific examples.

6.9.1. RNA bacteriophages.

Bacteriophages such as R17, f2 or Qβ possess the simplest genetic systems so far discovered. The nucleic acid chain contains slightly more than 3300 nucleotides and is able to code for only about 1100 amino acids. R17 and f2 particles contain two types of protein; there are 180 molecules of the coat protein (CP) mol. wt 14 700 and one molecule of a second protein (the aggregating protein, A). The sequence of the 129 amino acids in the coat protein has been determined. The A protein is required for adsorption of the particles to their host bacteria. It has a molecular weight of 35 000 and contains about 320 amino acids. A third

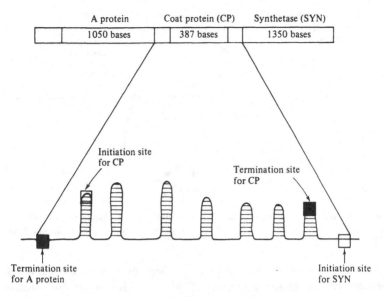

Fig. 6.10. The genome of R17 bacteriophage.

virus-specific protein is formed in infected cells. This is the RNA replicase (synthetase, SYN), an enzyme involved in the replication of the RNA chains. The order of the genes in R17 RNA has been established by nucleotide sequence and genetic studies. This provides the most detailed genetic map (fig. 6.10) at the nucleotide level of any virus so far studied and illustrates a number of important points regarding the control of translation of poly-cistronic virus RNA. Coat-protein molecules are produced in larger numbers than the others (approximately 10 CP : 1 A : 1 SYN). This regulation is carried out in two ways, one depending on the order of genes in the polycistronic messenger and the other involving the conformation of the RNA. Not all the nucleotides present in R17 RNA are translated into polypeptides and certain regions are involved with binding to ribosomes. It is these ribosome-binding sites that are intimately involved with the regulation of protein synthesis. Ribosomes have a very low affinity for the binding sites in the A gene whereas they bind readily to the CP gene, so that more coat protein can be formed than A protein, as shown in fig. 6.11. There are two important features regarding the ribosome-binding site of the SYN gene. First of all, this sequence of nucleotides is located in a helical region of the RNA so that it is not normally available for initiation purposes. Only after the synthesis of the CP protein does the helical region open and it becomes possible to start synthesis of the replicase enzyme. A second mechanism

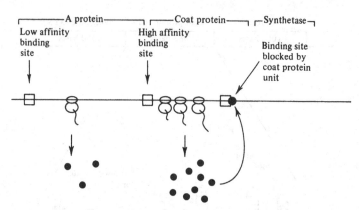

Fig. 6.11. Regulation of protein biosynthesis on R17 mRNA. (Modified from Watson, J. D. (1970). *The molecular biology of the gene.* W. A. Benjamin, Menlo Park, California.)

of control operates which can stop the synthesis of the replicase and which depends on the concentration of CP present. Five or six coat-protein molecules can bind specifically to the initiation sequences of the replicase gene and act as a repressor complex at the translational level. This is an example of how efficiently a simple virus genetic system can function – the structural protein possessing important control characteristics by stopping the synthesis of an enzyme which is not required in great quantity.

6.9.2. Picornaviruses (poliovirus). Polycistronic mRNA is seldom found in animal cells and the cellular genetic information is translated from monocistronic mRNA molecules. Infection of an animal cell with an RNA virus creates a problem. Because the virus is a polycistronic unit, it cannot be translated into individual polypeptides by the normal cellular ribosomal machinery. There appear to be two ways around this difficulty and different viruses seem to use particular mechanisms. In brief, each gene could give rise to a monocistronic mRNA molecule, one for each polypeptide, or, alternatively, the entire molecule could be translated as a monocistronic message to give a single very large precursor polypeptide.

Poliovirus appears to make use of the latter mechanism and this may result from the fact that the virus RNA itself can act as the mRNA and synthesis of virus nucleic acid is not required before the onset of virus-protein synthesis. Electron microscopy has shown that polysomes isolated from infected cells are much longer than cellular polysomes and their length is similar to that of virus RNA. Translation of a polycistronic mRNA in a genosphere which is only geared to translating monocistronic messages requires additional processes not necessary in the bacterial system described above. For example, the earliest proteins found in poliovirus infected cells are extremely large molecules. These large units, however, are soon cleaved by specific proteolytic enzymes to form the structural and functional polypeptide chains. Although this mechanism overcomes the failure of the mammalian ribosome system to recognise termination or re-initiation signals, it does not offer the same opportunity for the regulation of the relative amounts of different proteins formed as can occur with the R17 phage. In fact, the relative amounts of the virus-specific proteins detectable in poliovirus-infected cells remain the same during the major part of the growth cycle.

The study of the synthesis of virus-induced proteins in HeLa cells infected with poliovirus has allowed the genetic map of poliovirus to be established. When poliovirus infects HeLa cells, there is a rapid inhibition of synthesis of cellular protein and RNA. This host-cell repression occurs even when the replication of the virus is inhibited by guanidine (§ 6.7). Removal of guanidine from the infected cells allows the process of virus multiplication to commence in a highly synchronous manner and it becomes possible to detect newly formed virus-specific polypeptides. Fig. 6.12 shows the profile obtained after labelling infected cells with [^{14}C]amino acids and subsequent electrophoresis of the cell extracts on SDS–acrylamide gels. Use of inhibitors of proteolytic enzymes or amino acid analogues can cause the accumulation of the large polypeptides. Some temperature-sensitive mutants of poliovirus have been shown to cause the accumulation of the large polypeptides when infected cultures are maintained at non-permissive temperatures, for example at 39.5 °C instead of 37 °C. Evidence that a precursor–product relationship exists between the large and smaller polypeptides is based mainly on analysis of tryptic-digest finger prints of the various polypeptides involved and on kinetic studies of their biosynthesis. A recent important technique permits a direct analysis of the order of genes in a polycistronic messenger which contains only one initiation site as is found in picornaviruses. The drug **pactamycin** is known to inhibit selectively initiation of protein synthesis but does not prevent the completion of chains. There is therefore a transient period of protein synthesis after addition of pactamycin during which nascent proteins are completed and released. Hence, by addition of radioactive amino acids after pactamycin, proteins coded by the 3'-end of the mRNA will be more heavily labelled than those formed at the 5'-end. Another drug, **emetine**, blocks the nascent polypeptides on the ribosomes. Hence, very short pulses of radioactive amino acids

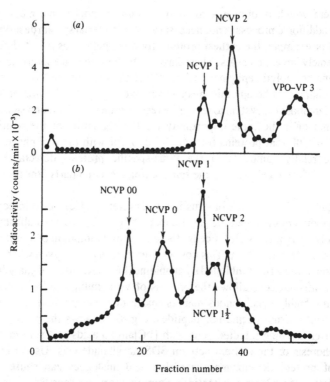

Fig. 6.12. Polyacrylamide gel electrophoresis of polypeptides made in poliovirus infected cells. (*a*) Without amino-acid analogues; (*b*) With amino-acid analogues. (From Jacobson, A. S., Asso, J. & Baltimore, D. (1970). *J. mol. Biol.* **49**, 657–99.)

followed by treatment of the infected cells with emetine permit the release of only these polypeptides labelled at the C-terminal end, that is, in the region translated from the 3'-end of the mRNA. Measurement of the amount of radioactivity incorporated into protein components after addition of these inhibitors makes it possible to deduce which of the polypeptides are translated first and allows an accurate determination of the gene order. Fig. 6.13 shows the sequence of events that follow the synthesis of the initial large polyprotein which is translated from the whole virus genome. This primary product called NCVP00 does not exist under normal conditions of infection as it is cleaved by proteolytic enzymes before it is completely translated. Cleavage of the primary product takes place in three steps. The first cleavage yields three fragments, one of which (NCVP1) is the immediate precursor of the structural polypeptides, the other probably being involved with

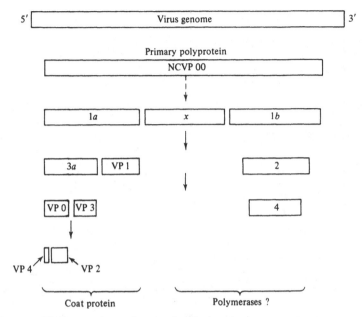

Fig. 6.13. The proposed genetic map of poliovirus.

replicase activity. The second cleavage leads to the production of the three capsid proteins VP 0, VP 1 and VP 3, and the third cleavage occurs after the formation of the pro-capsid to form VP 2 and VP 4 as mentioned on §5.11. It is of interest that the capsid proteins are formed on the 5'-end of the RNA whereas the replicase enzymes are produced at the 3'-end, that is, the region where the binding sites for polymerases are found.

Protein synthesis involving a sequence of post-translational cleavage steps is a common feature of many viruses whose genome can act as an mRNA and provides a subtle method of circumventing the cell's inability to recognise internal initiation or termination signals.

6.9.3. Vesicular stomatitis virus (VSV) VSV provides a contrast to poliovirus in its mechanism of viral protein synthesis. The VSV RNA is longer than that of poliovirus, but is also a single strand and sediments at about 40 S. In contrast to poliovirus mRNA which is a single species, the mRNA isolated from VSV-specific polyribosomes has a bimodal distribution at 28 S and 13 S in sucrose gradients (fig. 6.14). The 13 S component contains at least six species of mRNA, the molecular weights of which are approximately the size required for coding for each of the virus-specific

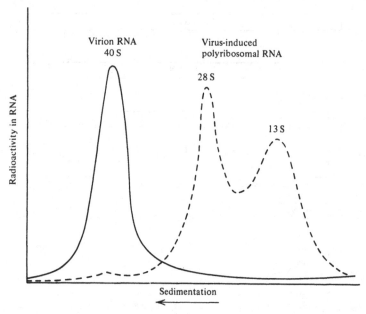

Fig. 6.14. A typical sucrose-gradient sedimentation experiment in which virus RNA and virus-induced RNA are compared. This type of experiment is often done by using two different radio-isotopes such as ¹⁴C (full line) and [¹³H]uridine (dashed line) and by use of double-labelling techniques compairions can be made on the single gradient.

proteins made in VSV-infected cells. These smaller RNA molecules are complementary to the virus RNA and have to be synthesised by infected cells before protein synthesis can begin. The result of this mechanism is that there can be a greater degree of control over the amounts of any particular protein produced than in the poliovirus system, since only monocistronic messages are translated. However, such a mechanism creates a problem which is not apparent in the picornavirus system. When virus RNA can be utilised as a messenger, the enzymes necessary for initiation of replication can be formed after infection. If RNA synthesis is required before translation can take place, however, some of the essential enzymes must either be present already in the cells or must be carried into them along with the infecting particle. The former is probably the case with DNA viruses which, soon after uncoating, can make use of the host cell's DNA-dependent RNA polymerase in order to synthesise virus-specific mRNA. As uninfected cells do not contain enzymes capable of replicating RNA molecules, viruses such as VSV must carry into the cell a replicase enzyme capable of initiating the processes involved in the production of mRNA molecules. Fig. 6.15

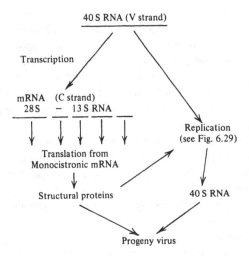

Fig. 6.15. A simplified scheme for the replication and translation of a Negative-stranded virus such as a vesicular stomatitis (VSV).

illustrates how the base sequences required for translation into virus protein are in the complementary strand and how this is transcribed into fragments so that each piece is a monocistronic mRNA. This means that the virus can use the normal cellular process for translation of the virus mRNA and no post-translational cleavage of large precursor proteins is necessary. Further-more, the use of the monocistronic messages permits regulation of the amounts of the various proteins formed. Regulation may be either at the transcriptional or translational levels, since the VSV proteins are not all produced in equimolar amounts, and may depend on the different amounts of mRNA produced or on the affinities of their ribosomal binding sites. With VSV, the nucleocapsid protein (N) is made in excess and forms a large cytoplasmic pool of nucleocapsid, while the membrane protein (M) is made in limiting amounts.

A similar mechanism operates with paramyxoviruses where segmented complementary strands are formed by a polymerase which is present in the virion.

6.9.4. Segmented genomes. An extreme situation has been adopted by certain viruses, such as orthomyxoviruses (influenza) and reoviruses in that their genomes are **segmented**. This means that instead of the virion containing one large RNA molecule there are a number of short molecules in each particle. Each fragment represents one gene of the virus. In some way, which is not yet understood, the individual segments can be organised into one unit so that each particle contains only a single copy of each gene.

Influenza is a single-stranded RNA virus and transcription of the virion RNA is required before protein synthesis can occur; thus, influenza has similar characteristics to paramyxovirus and rhabdovirus described in § 6.9.3 in that the complementary RNA (cRNA) strand acts as the mRNA. These groups of viruses have been referred to as **negative strand viruses**, the mRNA being defined as the +strand.

Reoviruses contain double-stranded RNA and also transcriptase molecules. The RNA is in ten segments, each corresponding to the size of one gene. At least ten transcriptase molecules are present which transcribe their RNAs into 10 mRNAs equivalent in size to the genome segments, and to the molecular weight of the newly synthesised polypeptides. mRNA is translated directly and there is no post-translational cleavage of precursor proteins. The amounts of proteins produced by the different genes depend on their size, as the smaller segments are transcribed more often than the larger ones.

In these sections (6.9.2–6.9.4) we have seen the various mechanisms that RNA viruses have adopted to circumvent the problem of how a polycistronic system can be translated in a genosphere which is only able to support the translation of monocistronic mRNAs.

6.9.5. DNA viruses. After infection with most double-stranded DNA viruses, the host cell's transcriptases function to produce virus-specific mRNA. The production of virus-specified sigma factors which can alter the specificity of the host polymerase enzymes has already been described in § 6.8. This scheme allows the production of early and late mRNA molecules and provides a system where transcriptional control plays a major role in the regulation of the growth cycle. Transcription of bacteriophage DNA often gives rise to polycistronic mRNA, whereas in animal viruses the primary products of transcription are processed by cleavage into smaller molecules and addition of polyadenylic acid to the 3'-ends. With most DNA viruses, including papova, adeno and herpes viruses, transcription and processing of mRNA take place in the cell nucleus. mRNA then leaves the nucleus and becomes associated with the host ribosomes in the cytoplasm and is translated in the normal manner.

On the other hand, poxviruses replicate completely in the cytoplasm of the cell and are unique among DNA viruses in that there is a transcriptase in the virion. The poxvirus DNA is partially transcribed into mRNA even before the uncoating process is complete, is released from the 'core' particles after polyadenylation of the 3'-ends and subsequently attaches to ribosomes.

In general, therefore, the transcription and processing of mRNA of DNA viruses follows closely the mechanisms used by non-infected cells, with modification, or replacement, of host-cell polymerases effecting a temporal control on the transcription of different groups of mRNAs.

6.10. Replication of DNA viruses

It is when we consider the mechanism of replication of virus nucleic acid that we see the real difference between viruses and cells. All kinds of cells, bacterial, animal or plant, undergo **binary division** which results from the

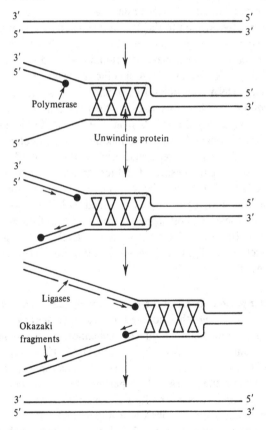

Fig. 6.16. A schematic representation of the essential features of DNA replication.

duplication of the genetic material. Each period of gene duplication in cells is separated by a relatively prolonged period of gene function during which metabolic processes are carried out. In general, viruses lack the control mechanisms which restrict replication of their nucleic acids to only one cycle and result in a period of continuous nucleic acid synthesis. The overall effect is a process more akin to amplification than to the replication of cellular genes. The process, however, depends on the same principles that operate in cells and always involves the phenomenon of complementary base pairing between precursor nucleotides and complementary template strands. In fact, much of our knowledge to date on the mechanism of DNA replication has been derived from studies on bacteriophage DNA; a number of enzymes are involved, such as polymerases, ligases, nucleases and unwindases. In brief, DNA polymerases extend polynucleotide strands in the 5' to 3' direction by addition of 5'-nucleotides on to the 3'-end (OH-end) of the primer strand. Ligases seal single-strand nicks in duplex DNA molecules by forming a phosphodiester bond between the 3'(–OH)-end and the 5'-phosphate on the other side of the nick. Certain DNA-specific nucleases play an important role by cleaving DNA molecules to specific sizes; sometimes these are structural proteins of the virus capsid. Also, certain exonucleases can selectively remove terminal residues from the 3'-ends to yield cohesive or sticky ends. Proteins have been found in some phage-infected cells which are able to unwind the DNA helix just prior to the point of growth, thus permitting synthesis of the nascent daughter strands to progress along the template. As DNA polymerases always proceed along a template in the 3' to 5' direction, the copying of duplex-stranded molecules has been found to involve the synthesis of short segments, referred to as Okazaki fragments, each of which is formed on each strand alternatively and then the pieces are joined together by ligases. A general outline of DNA replication is illustrated in fig. 6.16.

6.10.1. Circular permutations. The T-bacteriophages are linear duplex molecules which replicate using either modified host polymerases or virus-specified polymerases. The immediate products of replication are concatenated molecules up to four times the length of the parental DNA. These large structures may be formed by end-to-end joining of linear daughter replicas or by a cyclic process, called the rolling circle, which permits the continuous replication of a circular template to produce multi-unit structures. The outlines of these schemes of replication are described in figs. 6.17 and 6.18.

The formation of large concatenated molecules helps to explain many of

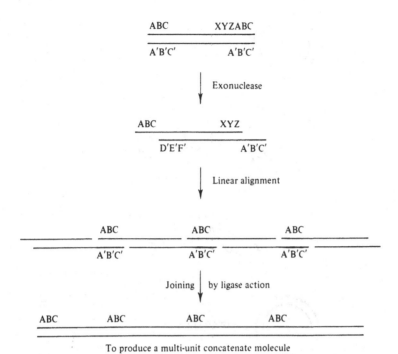

Fig. 6.17. A possible mechanism for the production of multi-unit strands of DNA. This use of **terminal redundancy** is probably involved in the replication of certain bacteriophages.

the structural and genetic properties of bacteriophages and the genetic map appears to be circular. The immediate precursor of virion DNA is a long multi-unit molecule which is cleaved by nucleases into a unit length prior to or during encapsidation. During replication of T4, DNA molecules which possess terminally redundant sequences such as ABC...XYZABC are formed. In fact, in a population of T4-phage all the DNA molecules do not start with the same sequence and, even though they are linear structures, genetic analysis has demonstrated that the genetic map is circular. This situation can arise by cleavage of the concatenate strands into units of the correct length to be packaged into phage heads. In fact, excision of DNA after encapsidation could mean that the head will take more than a single genome unit and hence lead to a collection of circularly permuted sequences (fig. 6.19).

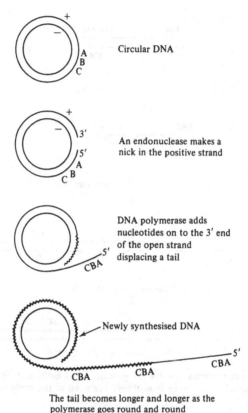

Circular DNA

An endonuclease makes a
nick in the positive strand

DNA polymerase adds
nucleotides on to the 3' end
of the open strand
displacing a tail

Newly synthesised DNA

The tail becomes longer and longer as the
polymerase goes round and round
the circular template (− strand).
A greater-than-unit-length DNA is formed

Fig. 6.18. **The Rolling Circle.** (Modified from Watson, J. D. (1970). *The molecular
biology of the gene.* W. A. Benjamin, Menlo Park, California.)

6.10.2. The rolling circle. This mechanism helps to explain a number of
features of DNA replication, even for DNA which is linear, like Lambda
phage, or single-stranded, like φX174.

Lambda (λ) phage DNA is a linear molecule with short terminal regions
of single-stranded complementary cohesive ends. There is an enzyme,
exonuclease III, in *E. coli* which can remove nucleotides stepwise from the
3'-ends of double-stranded DNA, thus producing molecules which possess
a 5'-ended single-stranded faction as shown in fig. 6.20. After infection, the
cohesive ends of the λ phage hybridise and a completely covalent circle is

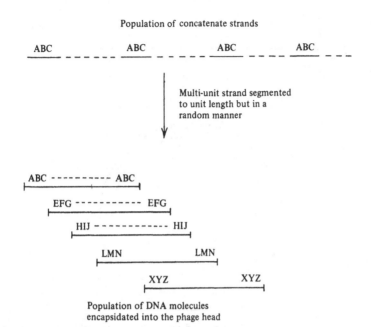

Fig. 6.19. The maturation of T4-phage DNA.

formed by ligase activity. Replication of the parental circle takes place to produce about 20 circles per cell by a process which starts at an initiation point and progresses bidirectionally from the origin to produce a theta (θ)-like structure which has been observed by electron microscopy. The second stage of replication involves a rolling circle mechanism in which one strand of the circle is nicked and the 3'-end of the nicked strand acts as a primer for elongation of the chain, using the closed circle as a template. This scheme also allows the production of large concatenated structures which are cleaved into unit lengths during the maturation process. One of the head proteins is able to excise the terminal sequences to form the cohesive ends which permit the formation of circles following the next infection.

A single-stranded DNA virus, such as ϕX174, also replicates by a rolling circle mechanism. Firstly, a double-stranded circular form is made by host-cell enzymes and acts as a **replicative form (RF)** during the asymmetric synthesis of the new progeny DNA strands involving virus coded protein. In contrast to the λ phage, however, the formation of the complementary strand is suppressed.

In general, replication of DNA viruses makes use of similar enzymes to

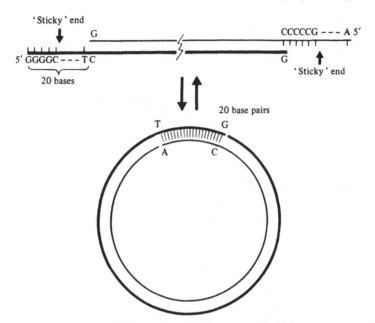

Fig. 6.20. Interconversion of the linear and circular forms of λ DNA. (From Watson, J. D. (1970). *The molecular biology of the gene*, W. A. Benjamin, Menlo Park, California.)

these involved in replication of cellular DNA and a common feature is the presence of a circular structure at some stage during the replication process. The widespread presence of circular DNA molecules suggests that the relevant form for replication of viral DNA is the circle and the linear form may be an adaptation, for example in T-phages, for injection of the nucleic acid through the narrow tail. It is interesting to note that only bacteriophages with tails have linear DNA molecules, all the rest having circular genomes. The finding that the newly synthesised molecules are often longer than the ingoing parental strands can be readily understood by postulating the involvement of a cyclic mechanism.

6.10.3. Lysogeny and transduction. So far, only the mechanism of replication of viruses which cause lysis has been considered. As mentioned in §6.5.2, some viruses have other methods of infecting cells, changing the basic character of the host cell. The so-called temperate bacteriophages can infect cells and then seemingly disappear without killing the cells. The bacteria continue to replicate for many generations and are immune or resistant to superinfection by the same or a related phage. These bacteria, however, contain at least one complete copy of the virus genome since they can be induced to commence production of bacteriophage particles and eventually the cells lyse and die. The important point is that, once the bacterial cell is infected, it carries the genetic information required for phage synthesis through many generations of cell division. In other words, the phage genes are being replicated in the same way as the bacterial chromosome. The process is called **lysogeny** and illustrates the principle of what is involved when viruses enter into a 'peaceful co-existence' with a cell.

The λ bacteriophage has been studied extensively and its lysogenic effect will be described briefly. We have already seen that λ phage has a double-

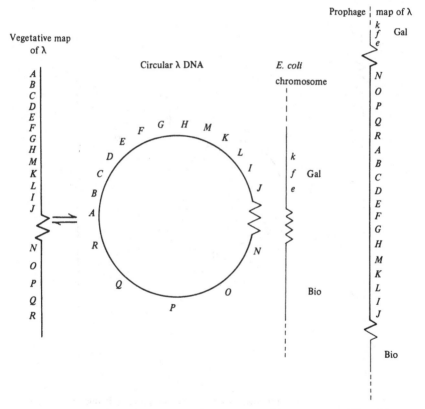

Fig. 6.21. The integration of λ DNA into the bacterial chromosome. (Redrawn from Hogness, D. S. (1966). *The structure and function of the DNA from bacteriophage Lambda.* In *Macromolecular Metabolism.* Boston, Little Brown & Co.)

stranded linear DNA which has cohesive ends (§6.10.2). On entering a bacterium, the phage DNA becomes a circle and occasionally this ring can be integrated into the bacterial chromosome. Integration takes place at a specific genetic site where there is a region of base-sequence homology. As shown in fig. 6.21, the phage circle is broken and integration is completed by crossing-over between the phage DNA and the bacterial chromosome. The phage is now said to be in the '**prophage**' state and some, though not all, of its genes are expressed. One particular gene produces a repressor protein which inactivates the remaining genes. This repressor also acts as the immunity substance which prevents superinfection. The cytoplasmic

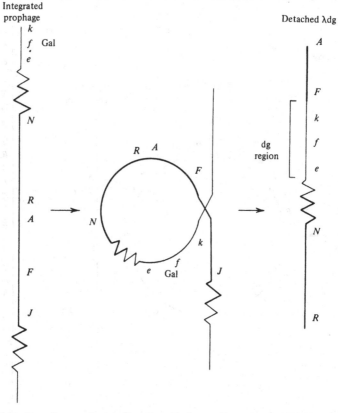

Fig. 6.22. Detachment of λ prophage from the host-cell genome. In this case, the loop of DNA was formed in such a way that not all the viral genome was included, but instead was replaced by cellular genome. The viral genome, therefore, is defective and carries the cellular genes of the galactose region. (Redrawn from Hogness, 1966).

concentration of the repressor protein is critical for maintaining the prophage state and, therefore, events which lower its concentration result in the complete expression of the virus genome. An interesting example of this occurs when a lysogenic male bacterium conjugates with a non-lysogenic female cell and carries into this cytoplasm the prophage DNA. Since no cytoplasm from the male cell is exchanged, the concentration of repressor protein is suddenly decreased, resulting in the rapid induction of phage multiplication and subsequent cell lysis and death.

No exactly analogous situation is found in animal cells, although certain aspects are seen in the integration of the DNA of oncogenic viruses into the host cell's chromosomes as occurs with polyoma virus.

The transition from a prophage state to a lytic infection involves the detachment of the phage DNA from the bacterial chromosome, thus re-establishing a circular structure. This is effected by the formation of a loop structure and the scission and subsequent relinking of the free ends. Sometimes the loop forms in such a way that a neighbouring cell gene is removed along with the virus DNA and is replicated during the synthesis of the new virus particles. It is known that λ phage is inserted into a bacterial chromosome between the galactose operon and the biotin cistron and, therefore, it is often found that the 'gal' cistron is removed and is present in the released phage particles. The phenomenon is known as specialised transduction (fig. 6.22).

A second type of transduction, known as generalised transduction, occurs when pieces of bacterial DNA become incorporated into the bacteriophage heads. Of course, such particles are non-infectious, but have been very useful in genetic studies.

6.11. Replication of RNA viruses

An intriguing situation arises when we consider the infection of cells by RNA viruses. The concept of the genosphere is particularly valuable here, since it defines very precisely the environment in which an invading RNA molecule finds itself. Spiegelman has written at length on this general problem in terms of a RNA genome invading a DNA-dominated universe and I suggest that his ideas are crystallised by the concept of the genosphere, the immediate environment created by the function of DNA genes. Such a genosphere appears to have no need for enzymes which can replicate RNA molecules, since all types of cell RNA are probably transcribed from DNA templates. Hence, when a RNA virus enters a cell, events must take place to modify, or convert, the DNA genosphere into a RNA genosphere. We have already seen that the first steps in this process may involve inhibition of cell DNA function. This in itself does not explain how RNA molecules can be replicated and until about 1962–3 our ideas were dominated by an erroneous interpretation of the Central Dogma which stated that genetic information was always passed from DNA to RNA and never vice versa. During the last ten years, however, many new data have shown that RNA viruses can successfully take over a DNA genosphere in a number of different ways, none of which contradict, in principle, the essential features of the Central Dogma and which even extend the concept and make it much more comprehensive. The basic tenet that amino-acid sequences are never, 'and probably cannot' be, translated back into nucleotide sequences remains unchallenged.

In brief, there appear to be two main mechanisms for the replication of RNA viruses. Some viruses code for an enzyme called 'replicase', which can replicate the RNA strands directly without involving DNA. Others can transcribe the base sequences of the viral RNA into DNA molecules which at some later stage can be used to transcribe the sequences back into virus RNA. These enzymes are called 'reverse transcriptases' (see p. 122).

Most RNA viruses contain single-stranded molecules and hence the essential problem of replication is to know how a given sequence is copied into the same sequence, for example, ACGUACG to ACGUACG. A possible mechanism may involve a direct copying of the strands rather than the involvement of complementary sequences. The information available from a wide range of viruses supports strongly the view that base-pairing between A and U and C and G is the key to replication, hence the fundamental principles which are found with DNA apply.

Our understanding of the replication of RNA viruses stems from two developments in the early 1960s. These are the discovery of RNA bacteriophages and the finding that picornaviruses are able to replicate in animal cells in which DNA synthesis and transcription into cellular RNA molecules are inhibited by actinomycin D (fig. 6.23). This drug has probably been used more often than any other in the elucidation of the mechanisms of RNA synthesis; without it our present knowledge of the biochemistry of RNA viruses would be considerably less advanced. Actinomycin D binds to DNA by interaction with deoxyguanine residues and prevents the cellular DNA-dependent RNA polymerase from functioning. Hence, cell RNA synthesis can be practically completely inhibited and the molecules produced after infection with a RNA virus can be studied, generally by allowing replication to proceed in the presence of radioactive precursors such as ^{32}P or ^{14}C or [^{3}H]uridine, which will specifically label only the newly formed RNA molecules.

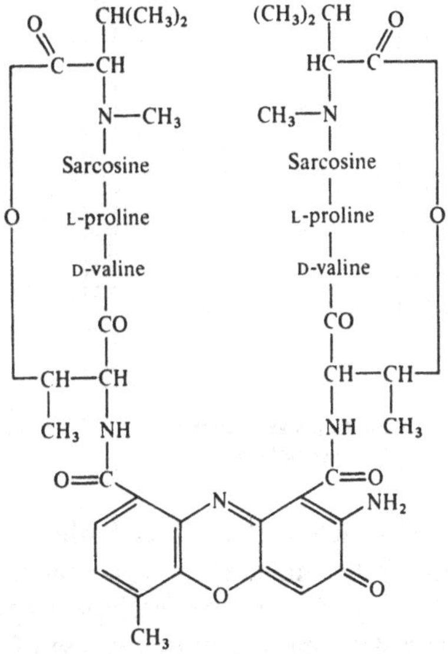

Fig. 6.23. The structure of actinomycin D.

The general protocol of experiments used in the study of RNA replication has been to inhibit cell RNA synthesis with actinomycin D and then to infect with the virus in the presence of radioactive precursors. At appropriate intervals, the RNA is isolated by treatment of the cells with phenol or detergents such as sodium dodecyl sulphate and the molecular sizes of the molecules are determined by sucrose-gradient analysis or acrylamide-gel electrophoresis. The presence of the virus-induced RNA can be followed by radioactive counting or measuring infectivity of the RNA fractions.

In 1963, Montagnier & Sanders made an important discovery by applying essentially the above technique to animal cells infected with encephalo-myocarditis (EMC) virus. They found that two types of RNA molecules, single-stranded virus RNA sedimenting at 35–37 S and a minor peak sedimenting at 18–20 S, were produced. The resistance of the 18 S component to degradation by ribonuclease under conditions which rapidly broke up cellular and 37 S virus-induced RNA was of great interest. Other tests indicated that this resistant RNA was a double-stranded form of the virus RNA and, by analogy with ϕX174, the new RNA was named **replicative form (RF)**. Similar double-stranded RNA molecules have been found extensively in both bacterial and animal cells after infection with many types of single-stranded RNA viruses (fig. 6.24).

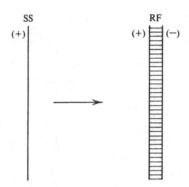

Fig. 6.24. The structure of the **Replicative Form** (RF), a Double-stranded RNA, which is resistant to degradation by ribonuclease.

A third type of RNA molecule was found which had characteristics intermediate between a double strand and a single strand. These intermediate molecules were partially sensitive to ribonuclease and mild treatment converted them to double-stranded structures. By use of the bacteriophage R17/*E. coli* system, Lodish & Zinder were able to label the infected cells for very short periods (15 s) and found label associated only with the intermediate peak. Extraction of RNA after a 'chase' period in non-radioactive medium, however, permitted the newly formed RNA to appear as a complete single strand, released from the double-stranded complex. This complex of growing single strands and double-stranded cores has been named the **replicative intermediate (RI)** as shown in fig. 6.25.

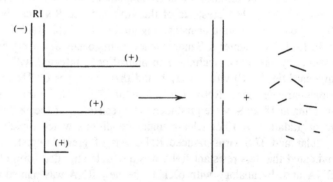

Fig. 6.25. The structure of the Replicative Intermediate (RI), a complex of double and single strands of RNA. The single strand parts are susceptible to ribonuclease attack.

At this point, it is important to realise that structures which are isolated from cells may not in fact be involved in cellular processes, especially when the extraction procedures may allow aggregation or complexing to take place. This is particularly the case when long regions of complementary base sequences are present that may readily anneal and form double-stranded molecules during extraction. The RF structures are therefore probably artefacts of extraction, but the RI fraction has real significance as it has been shown to have a number of growing tails which allows the asymmetric and preferential synthesis of single strands.

The most detailed description of the replication of an RNA has been provided by studies by Spiegelman and his collaborators on Qβ bacteriophage. The replicase enzyme was purified from infected *E. coli* cells in 1965 and shown to be completely specific for RNA from Qβ virus. Subsequently, complete replication of the virus nucleic acid was demonstrated under conditions in which only the enzyme, virus RNA and the mononucleotide precursors were present. The progeny molecules were infectious and were as biologically competent as those produced by an infected *E. coli* cell. This was proved by serial dilution of the replicase product and ensuring that none of the original virus RNA molecules were present as shown in fig. 6.26.

Weissmann and his colleagues have also studied intensively the synthesis of Qβ RNA and envisage that a sequence of events takes place as outlined in fig. 6.27. A replicase molecule binds to the 3′-end of a single-stranded virus RNA strand (V) and initiates synthesis of a single complementary strand (s) giving rise to a first-step replicating complex. It is possible that more than one enzyme molecule attaches and hence a number of complementary (c) strands are produced simultaneously. A second polymerase molecule attaches to the 3′-end of the complementary strand and synthesises a single-stranded virus (V) strand. During extraction procedures, involving, for example, phenol treatment, the complementary sequences may anneal to give the double-stranded complexes. These complexes are similar to those isolated from Qβ infected cells. The viral replicase of Qβ has been thoroughly investigated and is a complex of several different polypeptides. The core enzyme consists of four sub-units, α, β, γ and δ, the β unit being the only phage-coded polypeptide. The host sub-units are proteins involved with protein biosynthesis and have been shown to be the elongation factors (EF, Tu and Ts). A central problem to the replication of single-stranded RNA viruses, such as Qβ, is that the genome can also act as an mRNA. Hence, the replicase molecules and ribosomes both compete for the use of the genome and they move along the RNA strand in opposite directions. It appears that the replicases and ribosomes compete for an internal binding

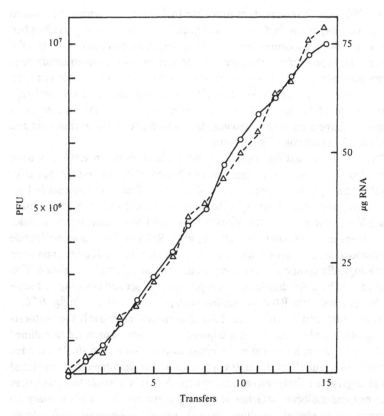

Fig. 6.26. In vitro RNA synthesis and formation of infectious units in a serial transfer experiment. Tubes containing RNA-replicating enzymes and cofactors were set up, and viral RNA was added to the first tube. After 40 min, an aliquot was transferred to the second tube. The transfers were repeated as indicated. All tubes were assayed for RNA content and for infectivity. The accumulated amounts are plotted here against the number of transfers. --△--, infectivity, —O—, RNA. (Redrawn from Spiegelman, S., Haruna, I., Holland, I. B., Beaudrean, G. & Mills, D. (1965). *Proc. Nat. Acad. Sci.* **54**, 919–27.)

site about half-way along the genome, and hence once attached the replicase can prevent further ribosomes from binding. The tertiary structure of the RNA ensures that the replicase is close to the 3′-end of the molecule, so that, when the last of the ribosomes have reached the 3′-end, the replicase can attach to the initiation site and commence replication. The replication of animal viruses, such as poliovirus, also involves the formation of replicative intermediate, but little is known about the detailed structure of the

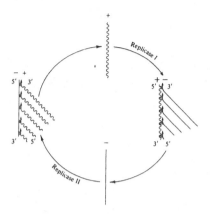

Fig. 6.27. Replication of bacteriophage Qβ.

polymerases. Animal virus polymerases are present in low amounts and have been difficult to purify. There are probably two polymerases involved in the replication of picornaviruses, one coded for by the 3'-end of the virus genome and this is considered to be the enzyme responsible for transcription of c-strands from the VRNA strand. Polymerase II is thought to be coded by the middle section of the genome and transcribes the new progeny VRNA from the c-strand template. An important feature of picornavirus RNA is the presence of a poly A tract at the 3'-end. This is essential for infectivity of the RNA and may act as a binding site for the polymerase (fig. 6.28).

Although models of RNA replicating structures are generally considered to be linear structures, some evidence favours a cyclic mechanism. Brown & Martin proposed in 1965 that FMDV replicated by a cyclic displacement mechanism, in order to account for observations that single strands of newly synthesised RNA appeared to be longer than the ingoing parental RNA (fig. 6.29). More recently, Agol has isolated circular double-stranded RNA molecules from EMC-infected cells and the presence of poly A and poly U tracts in the RI component suggests a possible way in which a cyclic intermediate may be formed. However, the question of circularity in RNA replication is still an open question and may only be resolved when we understand more fully the interaction between RNA and protein in the replicating complexes. In fact, it is very unlikely that virus RNA exists as a naked nucleic acid in animal cells, but soon after infection it becomes associated with host protein to form ribonucleoprotein (RNP), similar to mRNP. In viruses which contain a nucleocapsid, like paramyxoviruses, the RNP appears to be the active component in both replication and transcription

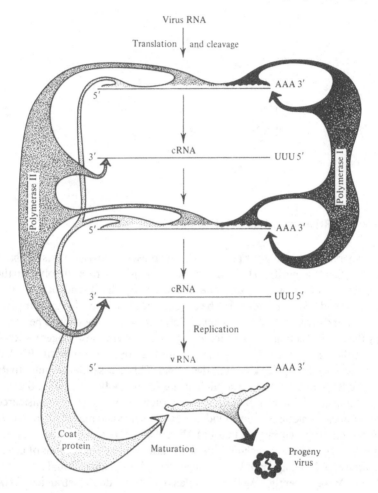

Fig. 6.28. Possible scheme for the replication of a picornavirus. For the sake of clarity the two polymerase activities required (I and II) have been located on different regions of the genome but the evidence for this is as yet circumstantial.

processes and a balance between these may be maintained by the availability of the capsid protein.

As mentioned in §6.9.4 the group of viruses known as negative-stranded viruses require to produce monocistronic mRNA as well as intact VRNA. In this case, the polymerase must be able to recognise internal termination sequences during the production of mRNA but be able to read through these

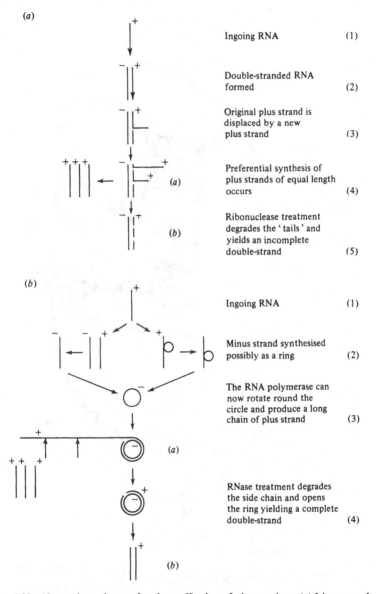

(a)

Ingoing RNA (1)

Double-stranded RNA formed (2)

Original plus strand is displaced by a new plus strand (3)

Preferential synthesis of plus strands of equal length occurs (4)

Ribonuclease treatment degrades the 'tails' and yields an incomplete double-strand (5)

(b)

Ingoing RNA (1)

Minus strand synthesised possibly as a ring (2)

The RNA polymerase can now rotate round the circle and produce a long chain of plus strand (3)

RNase treatment degrades the side chain and opens the ring yielding a complete double-strand (4)

Fig. 6.29. Alternative schemes for the replication of picornavirus. (a) Linear mechanism; (b) circular displacement mechanism. (From Brown, F. & Martin, S. J. (1965). *Nature, London,* **208**, 861–3.)

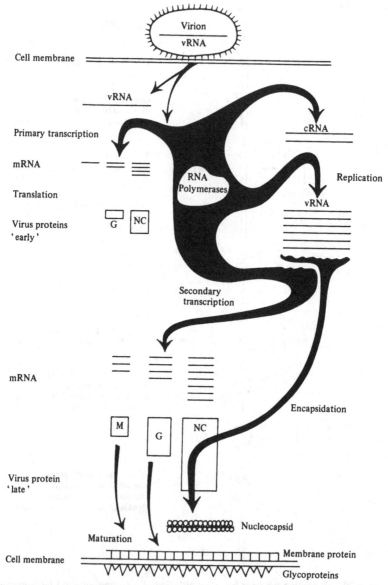

Fig. 6.30. A possible scheme for the replication of paramyxoviruses.

signals for the production of the intact complementary (+) strand which is utilised as a template for the synthesis of progeny virus RNA. A possible scheme for the replication and transcription of this type of virus is outlined in fig. 6.30.

6.12. Oncogenic viruses

Viruses that are capable of inducing either benign or malignant neoplasms are called **oncogenic**. They occur in both of the sub-phyla of viruses, deoxyviria and riboviria. Classification of viruses in general, however, is no longer based on host response but primarily on the properties of the virus particle and the DNA and RNA oncogenic viruses will be considered separately.

6.12.1. DNA oncogenic viruses. Viruses which cause tumours are widely distributed among the DNA virus families. In their morphology and DNA content, they cover a wide spectrum ranging from the large DNA-rich poxviruses to the small polyoma and parvoviruses. Viruses such as molluscum contagiosum and papilloma cause warts in humans.

The mechanism of transformation by DNA viruses is best understood by analogy with the phenomenon of lysogeny described in § 6.10.4. When a cell is infected with polyoma virus, for example, two courses are open; the virus can multiply and cause lysis of the cell with production of a large number of new particles or the infective virus can disappear. In the latter case, the cells are transformed and possess certain characteristics not present in non-infected cells. It is thought that the DNA of the virus is integrated into the chromosomes of the cell and that some, but not all, of the virus genes can function. Evidence for the presence and function of the virus genome came from the discovery that polyoma virus induces a specific new **transplantation** antigen in the transformed cell. Direct evidence for the integration of virus nucleic acid into cellular DNA can be obtained by hybridisation experiments. Radioactively labelled RNA has been isolated from transformed cells and tested for its ability to hybridise with viral DNA that had been denatured and firmly attached to cellulose filters. RNA molecules which have a base sequence complementary to the virus DNA strands will be held in the filter. This technique provides a very sensitive method for detecting the presence of minute amounts of genetic material that have an extracellular origin.

Unlike lysogenic bacteria, the transition to a lytic state does not always exist in animal virus systems. SV40 can be induced to produce new infective particles by fusing transformed cells to non-transformed cells. This indicates that the complete genome of SV40 is retained in the transformed cells for many cell generations; the situation is similar to the induction of lysis during the mating of bacterial cells (see p. 111). Not all animal viruses behave like this and oncogenic adenoviruses and polyoma virus cannot be recovered from transformed cells, possibly because not all the virus genome is present.

In summary, DNA viruses probably cause transformation of cells by the integration of all or part of the virus genome into the host-cell chromosome. In some ways this can be looked on as a method by which the cell defends itself against a fatal attack by a lytic virus. By incorporating the virus into the chromosome, the cell prevents lysis and inevitable death. The 'minor' alterations effected by the additional genomes are small sacrifice for 'life' and the fact that some of these become malignant in animals may be a result of a deficiency in the individual's immunological response to the altered cells. In the vast majority of cases, the initial clone of transformed cells may be rejected by immunological mechanisms. For example, polyoma virus is widely distributed among populations of wild mice and appears to cause no serious illness or cancer and yet readily causes tumours in immunologically incompetent animals.

6.12.2. RNA oncogenic viruses. RNA oncogenic viruses differ in many ways from the DNA viruses described above. One of the most striking differences is the ability of cells transformed by RNA viruses to continue to produce infectious virus particles. Also, the RNA viruses causing tumours all belong to the same morphological group which has been called **retroviruses.** These viruses have certain similar structural features to myxoviruses; however, they do not cause visible cytopathic effects and the production of virus particles becomes a new excretory function of the transformed cells, which they perform without dying in the process. In mammalian and avian species, many of the viruses are capable of inducing leukaemia and similar diseases and others cause solid tumours such as Rous sarcoma, mouse sarcoma and the mouse mammary tumour.

The mechanism of replication of these viruses poses intriguing problems for the biochemical virologist, since it could mean that RNA genomes persist in mammalian cells for generations. Howard Temin, however, realised as long ago as the early 1960s that the biology of RNA tumour viruses could be best explained by involving an enzyme capable of transcribing the base sequence of RNA into DNA molecules with the subsequent integration of the new virus-induced DNA into the host-cell chromosome. At the time, this idea was contrary to the canons of molecular biology, but he persisted with his hypothesis that RNA tumour viruses could enter a 'provirus' state in animal cells analogous to the 'prophage' state described for lysogenic bacteriophages. In 1970, both Temin and Baltimore, working independently, found that there was an enzyme present in the Rous sarcoma virus (RSV) particles which could synthesise DNA using the RSV RNA as a template. This important discovery opened up new horizons not only to the field of tumour virology but to our whole understanding of the interaction between nucleic acids.

Once again, hybridisation experiments have been of immense importance in establishing the true significance of the existence of these so-called 'reverse transcriptase' enzymes. It has been possible by this technique to prove that the DNA from cells, previously transformed by RSV, contains base sequences which are complementary to the RNA from RSV. Hybridisation also offers a method for demonstrating the presence of virus-related sequences in human malignant and 'normal' cells. Retroviruses have recently been recovered from a number of human leukaemias which resemble the oncoviruses associated with certain monkey sarcomas by both nucleic acid hybridisation and serological relatedness. However, as can be imagined, it is very difficult to prove directly that these viruses are the causative agents of the human cancer.

Once the virus genome is integrated, it is replicated at the same time as the cell DNA and some of the virus genes are expressed. The contribution of new proteins to the cell may effect a change to malignancy. It ought to be remembered, however, that the virus itself may not always be the cause of the malignant state; malignant cells may favour the integration of virus genomes which are carried merely as genetic passengers, contributing new characteristics but not themselves causing the malignancy. As always, there are two sides to the story. The fact that a cell integrates a virus genome prevents the virus from causing lysis and cell death. On the other hand, the ability of a virus to integrate with a cell allows the virus genome to continue to function and to be biological active (that is, alive), instead of being encapsulated and released as a 'dead' particle. The exact role of RNA tumour viruses in causing transformation is still not known, but there is an increasing body of evidence that suggests that the genome of these viruses (the virogene) is carried by all vertebrates and that their expression is induced following the changes that take place during or after tumour formation. For example, leukoviruses can be induced in clones of mouse embryo cells by treatment with carcinogens, such as 5-bromodeoxyuridine. These observations have led to the proposal by Todaro & Huebner of the **oncogene hypothesis** which emphasises the close association between viral and cellular genomes. In normal cells, the virus RNA genome (virogene–oncogene) is suppressed, but under certain conditions this can be partially or fully activated leading to tumour formation and the production of virus.

6.13. Maturation of virus particles

Some aspects of the maturation of viruses in relation to virus architecture have already been discussed and it is only necessary to re-state here the general principles involved. Simple viruses, containing only nucleic acid and protein, are formed by a self-assembly process. Two types are known. One (for example, TMV) requires the presence of the nucleic acid and there is a high degree of interaction between the capsid and the nucleic acid. The other mode of maturation appears to involve the assembly of empty particles which then become filled with nucleic acid. This appears to be confined to icosahedral viruses.

Two other types of maturation process take place which involve more complex steps. During the maturation of T-phages, the regulation of gene expression is intimately involved with the morphology of the complex particles, and the maturation of paramyxoviruses illustrates how enveloped viruses may be formed at the cell surface.

6.13.1. The assembly of T-bacteriophages. In the maturation of the T-even phages, at least 45 genes are involved and a stepwise process operates as shown in fig. 6.31. Two large sequences of genes are concerned with the production of head proteins. Formation of the heads requires the condensation of the complete virus DNA into a tightly packed ball-like form. Condensation may require an internal head protein and the protein shell is formed around the DNA. The tail plate is produced by a number of other genes and after assembly of the tail it binds spontaneously to the head. Five genes are required to form the fibres which bind to the base plate to form the complete phage. Even in this complex structure stepwise assembly of different components probably occurs spontaneously, since successful assembly of purified sub-components has been achieved in cell-free extracts.

The complete phage particles accumulate in the cell, which swells up by the intake of water due to changes in membrane permeability and literally explodes. The bursting of the cell, however, is an event controlled by a virus enzyme, lysozyme. This enzyme, which is attached to the tip of the tail fibres of the phage, was used for entry into the cell (see p. 77) and, at the completion of the replication cycle, when all the new virus particles have been matured, the same enzyme provides a means of escape from the devasted genosphere.

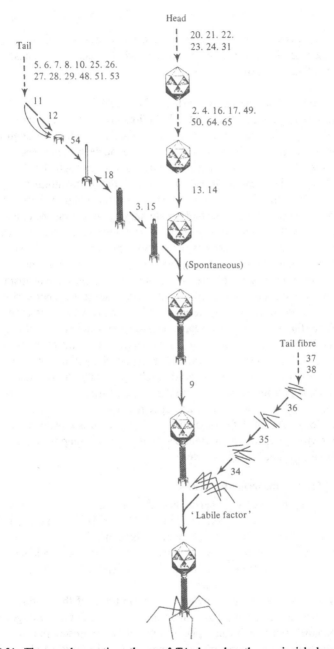

Fig. 6.31. The morphogenetic pathway of T4 phage has three principle branches leading independently to the formation of heads, tails and tail fibres, which then combine to form complete virus particles. The numbers refer to the gene product or products involved at each step. The solid portions of the arrows indicate the steps that have been shown to occur in extracts. (From Wood, W. B. & Edgar, R. S. (1967). Building a bacterial virus. *Readings from Scientific American. The molecular basis of Life.* W. H. Freeman & Co.)

Not all viruses are so well armed as these complex bacteriophages, but each type does have a precise mechanism which ensures its release from the redundant host cell.

6.13.2. Maturation of enveloped viruses. The nucleocapsid or core of complex viruses is generally surrounded by a lipid-containing membrane. This outer membrane(s) is produced by the nucleocapsid 'budding' through a cell membrane. Some particles pass through the nuclear membrane or the membrane of a cell vesicle before finally passing through the outer cell membrane. In these cases, viruses may have a double membrane.

The mechanism of maturation is best known for myxo- and paramyxoviruses and these will be described briefly although of course the details are not applicable to other virus groups. With influenza virus and with the paramyxoviruses such as measles, a two-stage process occurs. The nucleocapsid is formed either in the cytoplasm or at the nuclear membrane. Electron micrographs of infected cells show the naked nucleocapsid filaments in the cytoplasm and they appear to align along the inner surface of the cell membrane. During the period of infection many biochemical events take place, the details of which have not yet been elucidated, but the final outcome is the modification of the cell membrane. Proteins of the host cell appear to be replaced by glycoproteins specified by the virus and then somehow the virus core is extruded through this membrane. A small packet of membrane surrounds the core and the mature particles bud off. Whereas the proteins of the membranes are specified by the virus, the host cell appears to be the chief donor of the lipid of the virus envelope, since if the same virus is grown in different cell types its lipid composition is found to be similar to that of the host cell.

6.14. Defective interfering particles

Many viruses possess a property, first discovered by Von Magnus with influenza virus, which is known as **autointerference**. This occurs when virus is passed in tissue culture at high multiplicities of infection and results in a drop in the titre of the released virus. Evidence is also available that a rhythmic process operates in that the titres of released virus increase and decrease with continuous passage.

These characteristics are now understood in terms of the production of **defective interfering particles (DI)** which contain only part of the virus genome (**sub-genomic**), but which can replicate when they are present in cells along with infectious virus. In general, the small size of the DI RNA allows it to replicate more rapidly than the intact genome and hence competes strongly

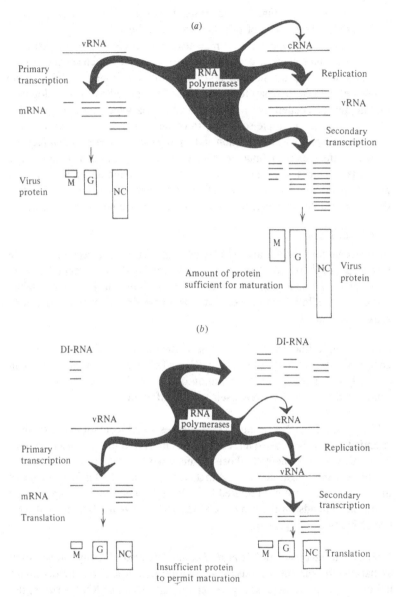

Fig. 6.32. The effect of **defective interfering particles** (DI) on the replication of standard virus.

for the polymerases. Fig. 6.32 illustrates how the presence of DI particles can decrease the amount of infectious virus produced.

Defective interfering particles may also be involved in the establishment of persistent infections, especially with certain viruses, such as rabies, VSV and measles. The mechanism of this is still not understood but is an important area in virology today as it may be related to the development of long-term degenerative neurological diseases which may have a viral etiology, such as multiple sclerosis. In certain conditions, DI particles can alter the lytic pathway of infection and rapidly establish persistent infections in which the cells continue to survive but carry in them considerable amounts of virus antigens. In an animal, the outcome of this type of persistent infection may trigger an 'autoimmune' response as the individual's immunological system attacks the cells which harbour the virus.

6.15. Conclusions

In this chapter, the various strategies that viruses can use to invade a foreign genosphere have been outlined. At first sight it may seem that there is no general underlying theme. Sufficient information is now available, however, to allow some generalisations to be drawn, which are briefly summarised here.

6.15.1. Simple riboviruses. The simplest situation is found in viruses which contain a single strand of infectious RNA. This means that viral RNA can be used as messenger for the initiation of virus-induced protein synthesis. With these viruses, the RNA itself will be infectious.

6.15.2. Infectious DNA. Simple DNA viruses which contain either double-stranded or single-stranded infectious DNA must be initially transcribed into early mRNA components by host-cell enzymes.

The principle illustrated here is that, if the nucleic acid is infectious and the coat proteins are not required for infection, then the first step of virus synthesis depends on the use or production of active mRNA molecules by use of host-cell enzyme(s).

6.15.3. Non-infectious nucleic acid. The nucleic acid of many viruses such as paramyxovirus, myxovirus, rhabdovirus, poxvirus and reovirus is not infectious. In all these cases, whether they are DNA or RNA viruses, the virion takes into the cell an essential enzyme. In the case of paramyxoviruses, a novel ($RNA \rightarrow RNA$) transcriptase appears to be required for the synthesis of 'active' mRNA molecules, and with poxvirus a viral-specific ($DNA \rightarrow RNA$) transcriptase is taken into the cell.

6.15.4. RNA–DNA viruses. We have also seen numerous examples of how both DNA virus and RNA viruses can become integrated into the DNA of the host cell. The discovery that a transfer of information from RNA → DNA is possible means that a process akin to what occurs with lysogenic bacteriophages may also take place in higher organisms, even if the nucleic acid of the virus particle is RNA.

6.15.5. Cell-to-cell migration. Many viruses have the property of causing syncytia formation, which follows the disruption of intercellular membranes and the production of multinucleated cells. This is associated with the property of cell fusion which can be caused by some membrane viruses and means that particles can readily pass directly from one cell to another. Such a mechanism of infection has obvious advantages in animals when a virus infection can spread without being attached by the circulating antibodies and lymphocytes in the blood stream.

6.15.6. Interaction with sub-genomic (DI) particles. During the replication of many viruses, especially at a high multiplicity of infection, there appears to be the accumulation of particles containing sub-genomic RNA. The presence of such particles can cause interference with the normal replication of the virus as they can compete for the available polymerases that are present. It is also possible that these defective particles participate in setting up persistent infections.

7 Viruses and the biosphere

7.1. Evolution

As we have seen, viruses may be considered as invaders of the genosphere, depending for their replication on the energy sources, the ribosomes and certain enzymes of their host cells. Viruses must therefore have evolved after the appearance of primitive cells, possibly as parasitic degradation products, and the infectious diseases they cause have provided powerful selective pressures on the evolution of the biosphere. Furthermore, the recent findings that both DNA and RNA viruses can become integrated into çells makes it possible that 'viruses' may have played an important role in the acquisition of increasing amounts of genetic material by primitive cells during the evolution of animals and plants. However, once complex communities in the biosphere had developed, the more sinister aspects of virus diseases soon became the predominant feature. In particular, Man's impact on the biosphere has offered viruses an unparalleled scope for world-wide distribution. Increasing urbanisation, intensive farming methods and swift and large-scale travel across the globe has made it imperative to be able to diagnose rapidly, to prevent and to cure virus diseases of organisms as different as trout, lettuce plants, bees and Man. In this chapter we shall consider some of the approaches to the therapy and prophylaxis of virus infections such as vaccination, chemotherapeutic agents and interferon. Also, we shall see that recent advances in molecular biology may lead to the development of genetic engineering technology, using viruses as tools, which may mimic an earlier role that viruses may have played in evolution.

7.2. Vaccines

Jenner, the discoverer of vaccination, was a country doctor in Gloucestershire and like many others was aware that farm workers who had contracted cowpox were immune to smallpox. His investigations showed that both natural and artificial infection of dairymaids and children with cowpox gave resistance to smallpox infection. The basic principles of vaccination were formulated on these early astute observations by Jenner and in fact, the vaccinia virus currently in use in laboratories around the world is by repute the linear descendant of the virus Jenner obtained from cows and dairymaids around the year 1796. The essence of Jenner's discovery was the demonstration that similar, but non-virulent viruses or antigens can stimulate the production of antibody after artificial exposure of an animal or person. The present era of successful control of numerous virus diseases can be attributed to the work done during the early 1950s by Salk and Sabin on the development of an efficient poliovaccine. These studies resulted from the discovery that polioviruses (there are three serologically distinct types) can grow in cultured monkey-kidney cells. Two approaches were made. Salk produced an inactivated vaccine for all three types of poliovirus by treatment of the virus with dilute formaldehyde at 37 °C and pH 7.0 for one week. These inactivated viruses, however, still retained adequate antigenicity to promote antibody formation when injected into animals or men. Sabin isolated three attenuated strains of poliovirus by multiple passages in a foreign host, most frequently tissue-culture cells. These strains lacked neurovirulence and have been successfully used as vaccines. The advantages of live vaccines are clear, since it is possible to administer them orally whereas an inactivated vaccine must be injected under sterile conditions, increasing the expense of production as well as administration. Nonetheless, the disadvantages of live vaccines are substantial and thorough investigations must be made to determine the genetic stability of the chosen strains before use. The genetic stability of the poliovirus vaccines has now been well established and the live vaccine is used predominantly. With live vaccines there is often the danger of the presence of other viruses which may be pathogenic, such as human tumour viruses, or that the attenuated virus, although not producing an acute illness, may lead to the development of long-term persistent infections which may not be detected until many years after the vaccination programme has started.

It is not always possible to isolate a genetically stable strain of virus which will not revert to a virulent form and here inactivated vaccines of some form. must be used. It should be remembered that most vaccines contain many components derived from the host cell as well as the relevant virus antigens. Often these irrelevant components can lead to side-effects such as the development of autoimmune encephalomyelitis which frequently follows repeated doses of rabies vaccine containing rabbit-brain tissue, or the hypersensitivity to penicillin which occurred in many recipients of the Salk vaccine for polio, as the antibiotic had been added to the growth media of the tissue cultures used in the production of the vaccine. These problems have led to the development of techniques to make vaccines consisting solely of the relevant immunogenic material, such as the capsid proteins of icosahedral viruses, like adenovirus, or the surface glycoprotein of influenza virus. These sub-unit vaccines, although preferable, are extremely expensive to produce and require a thorough knowledge of the structure of the viruses involved. In fact one of the long-term aims of current studies on the structure of viruses is to isolate the immunogenic components. At present our knowledge of the chemical nature of the antigenic sites is very limited and it is not yet possible for chemists to design synthetic antigens which will mimic the natural viral components. The continued interest and involvement of chemists in virology may well lead to the development of synthetic antigens for specific viruses as an important aspect of future immunisation programmes.

7.3. Chemotherapeutic agents

The discovery of penicillin and streptomycin sparked off a revolution in the chemical treatment of diseases caused by bacteria. This was because bacterial and mammalian cell metabolism differ sufficiently to allow the isolation of agents which would selectively affect bacterial cells, but have little, if any, effect on the infected animal. As we have seen, viruses depend on the biosynthetic processes of the host cell for their replication and this fact severely limits the possibilities for the successful and selective inhibition of virus multiplication. However, the list of substances that are able to inhibit viral replication is immense and many growth points in this field of research offer hope for the future, although to date only a few agents have had any practical value and in general chemotherapy has had a negligible effect on the control of virus diseases.

Specificity of drug action against a given virus could be achieved by pinpointing a particular metabolic step or reaction which is essential for virus replication but not necessary for the life of the cell. For example, if a specific inhibitor could be found for the replicase enzyme produced by RNA viruses many serious virus diseases could be controlled. Some compounds that inhibit different stages of the growth cycle of the virus will now be described. Their structures are shown in fig. 7.1.

The attachment and penetration of virus into cells are highly specific processes and offer obvious points of attack. Amantadine (L-adamantamine hydrochloride), one of the few antiviral drugs currently available on prescription, prevents the penetration of myxoviruses e.g. influenza virus. The mechanism of action is not known.

Many agents which may affect nucleotide metabolism have been investigated and some of these are currently producing encouraging results. In particular, halogenated derivatives of deoxyuridine, such as 5-fluorouracil,

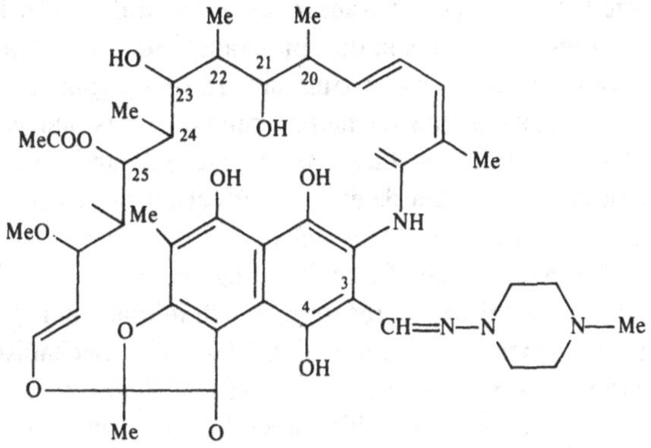

Rifampicin

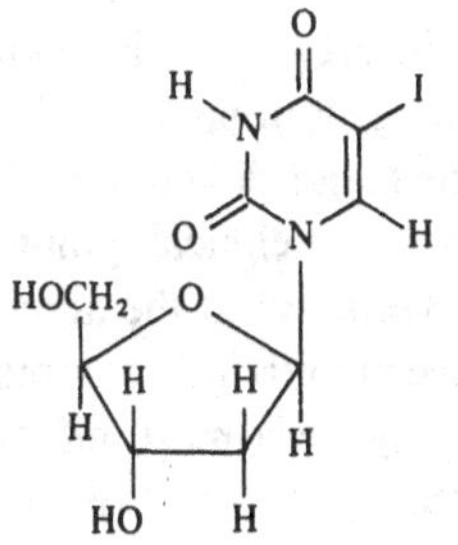

5-Iodo-2'-deoxyuridine

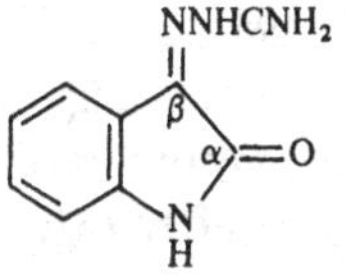

Isatin β-thiosemi-carbazone

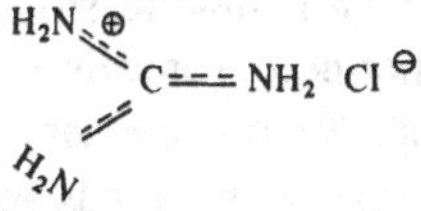

Guanidine HCl

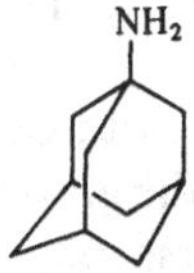

1-Amino adamantane hydrochloride

Fig. 7.1. Examples of chemotherapeutic agents against viruses.

5-bromouracil and 5-iodouracil, can interfere with the development of DNA viruses. These derivatives have two main types of action. 5-Fluorouracil deoxyriboside is inhibitory because its 5-monophosphate inhibits thymidylic acid synthetase, the enzyme which converts deoxyuridylic to thymidylic acid, and thus stops DNA synthesis by limiting the thymidine triphosphate. The bromo- and iodo- derivatives, which are sterically similar to thymidine, are metabolised by the cell and become incorporated into the newly formed viral DNA, in place of thymine. Although DNA synthesis can continue, the halogenated derivatives are transcribed into viral mRNA in an abnormal fashion and result in errors in the amino-acid sequence of the virus induced protein. New particles then fail to mature. These drugs are particularly useful for their inhibitory action on herpes simplex virus and vaccinia virus in surface lesions. However, their use for internal administration is severely limited, since they have a deleterious effect on normal cells that undergo rapid cell division such as blood cells.

One of the most successful antiviral agents so far tested in the field is *N*-methylisatin β-thiosemicarbazone, which inhibits multiplication of pox-virus but not other viruses containing DNA. The mechanism of inhibition is prevention of synthesis of late structural proteins by causing a defect in the late viral mRNA. The β-thiosemicarbazone drugs are of considerable practical use for the control of smallpox although the side-effects are considerable. *N*-Methylisatin β-thiosemicarbazone was successfully used as a prophylactic in preventing the spread of smallpox in an epidemic in Madras, India, in 1963, although it failed as a therapeutic agent in patients clearly suffering from smallpox. This illustrates one of the major difficulties of the chemotherapy of viral diseases. In most viral diseases the clinical symptoms only appear after extensive multiplication of the virus, when the adminis-tration of an inhibitor may be of little value. The use of antiviral compounds as prophylactic rather than therapeutic agents may be a more effective approach to the control of viral diseases, but immense dangers exist in the widespread and continuous use of agents for the prevention of infectious diseases, since the appearance of virulent drug-resistant mutants is highly probable.

There has been a great deal of interest recently in the antibiotic, **rifampicin**, which inhibits the growth of vaccinia virus. Furthermore, some derivatives inhibit the reverse transcriptases (RNA-dependent DNA polymerase), of the virions of several RNA tumour viruses. This group of compounds appears to have a good potential since molecules similar to rifampicin possess a common property of binding to DNA-dependent RNA polymerases, but small variations in the structure enable some derivatives to inhibit the reverse transcriptases found in oncogenic viruses. Also *N*-dimethyl-rifampicin has been shown to inhibit the growth of hamster cell lines transformed by polyoma virus under conditions where it had no inhibitory effect on the growth of two non-transformed hamster cell lines. Clearly these observations are of great interest, since they may offer a further means of therapeutic treatment of cancers which have a virus etiology.

A great stimulus for the search for antiviral agents came in 1957, when

Isaacs & Lindenmann showed that a substance, which they called **interferon**, was induced in cells in response to infection by certain viruses. Interferon gave the cells an immunity against infection by other types of viruses and more importantly could be successfully transferred to other cells making them also resistant to infection. In their experiments Isaacs and his colleagues found that when heat-inactivated influenza virus was allowed to adsorb to chicken embryo cells a product was released into the medium which showed antiviral activity against a range of other viruses, such as Newcastle disease virus, Sendai and vaccinia. The discovery of interferon produced a tremendous amount of excitement since it was shown that a number of viruses could induce its production even after they were in-activated by heat treatment or ultraviolet irradiation. The possibility of discovering a universal antiviral agent loomed before clinical science and a great deal of research ensued.

Interferon is the general name for a large group of substances which have a high degree of species specificity. For example, interferon produced by mouse cells protects other mouse cells, but gives little immunity to human and other cell types. In other words, the exact nature of the interferon produced does not depend on the nature of the inducing virus but rather on the origin of the infected cell. Many viruses can induce interferons although their ability as inducers may vary greatly among strains of the same virus. Of immense importance was the finding that double-stranded RNA components of viruses and also double-stranded synthetic nucleic acids could stimulate interferon production.

Interferons are small proteins with molecular weights in the range 20 000–40 000. However, the initial hope of obtaining interferon in sufficient quantity and purity for clinical use has now waned and current research is aimed at finding an efficient and non-toxic interferon inducer. Unfortunately, the most efficient interferon inducers, such as (Poly I . Poly C) are highly toxic even though in cell cultures they show marked inhibition of viruses. However, favourable results have been reported with double-stranded poly-ribonucleotides in the prophylaxis of a wide variety of experimental virus infections in animals. Also (Poly I . Poly C) has been shown to inhibit the growth of a variety of tumour cells.

The genetic information for interferon production appears to be contained in cellular DNA. Actinomycin D inhibits the synthesis of interferon and hence in some way the interferon inducer must be able to set in process the mechanism for the derepression of the cellular 'interferon' gene with the subsequent synthesis of interferon mRNA and its translation into the active protein. The nature of the stimulus that activates the cellular gene is unknown and it does not depend on the genetic expression of the infecting virus since either inactivated virus or synthetic double-stranded molecules can act as an inducer. In fact, recent findings suggest that the inducer does not need to enter the cell at all and the activation of the cellular gene may be a response to a change at the cell membrane initiated by the binding of the inducer to receptor sites.

Interferon does not interact directly with virus particles but inhibits the formation of complexes between virus mRNA and cellular ribosomes. The immediate action of interferon is to derepress a second cellular gene and permit the synthesis of a protein which has been named **translation inhibitory protein (TIP)**. TIP does not inhibit the translation of cellular mRNA but prevents the formation of polysomes containing viral-specific RNA.

7.4. Alpha and omega

As mentioned above (p. 130), viruses may have played an important role in evolution, not only by imposing severe selective pressures on susceptible host cells but also, by their ability to integrate with cellular DNA, they may have participated in the acquisition by cells of novel genetic material. Recently we have seen the emergence of techniques which may allow some viruses to be used advantageously. For example, viruses which specifically invade tumour cells could be used in the treatment of cancer and there have been some encouraging investigations on this topic. The use of viruses to eradicate insect pests would be a convenient way of dispensing with the need to use toxic chemicals in pest control and would be immensely important to agriculture and tropical medicine. Finally, the current understanding of the molecular biology of bacteriophages and viruses and the mechanisms of integration of foreign DNA into bacterial cells has opened up the possibility of genetic engineering. Of course the ability to insert genetic material of one species into another could have great benefit in many areas, such as the cure of metabolic diseases in Man or the creation of disease-resistant strains of important plants. Nonetheless, collaboration of Man with viruses is laden with profound ethical problems and hazards, such as the production of a virulent pathogen that could cause untold damage to the biosphere. Therefore, although our fight against these invaders of the genosphere will continue as long as life persists, we must ensure that our future technology does not permit them to win!

Finally, perhaps, in this microcosm of life we can see our own destiny. Will Mankind eventually escape from our devastated and decaying biosphere, by encapsidation and flight, to infect a greener and more plentiful planet?

Appendix   Virus classification

Phylum: *Vira*
Subphylum: *Deoxyvira*

Class	Family	Genus	Diseases and common viruses	Family cryptogram
Deoxyhelica	Poxviridae (enveloped)	Poxvirus / Molluscovirus	Smallpox / Benign human tumours	$\frac{D}{2} \frac{200}{6} \frac{X}{X} \frac{V}{0}$
Deoxyculica	Microviridae (naked)	φX174		$\frac{D}{1} \frac{1.6}{25} \frac{S}{S} \frac{B}{0}$
	Papovaviridae (naked)	Papilloma virus / Polyoma virus	Human warts / Mouse tumours	$\frac{D}{2} \frac{3\text{-}5}{7\text{-}13} \frac{S}{S} \frac{V}{\text{O, Di, Ac, Si}}$
	Adenoviridae (naked)	Adenovirus	Tonsillitis	$\frac{D}{2} \frac{20\text{-}29}{12\text{-}14} \frac{S}{S} \frac{V}{0}$
	Iridoviridae (naked)	Tipula iridescent virus	Insects	$\frac{D}{2} \frac{130}{15} \frac{S}{*} \frac{1,V}{*}$
Deoxykinala	Herpesviridae (enveloped)	Herpes virus	Cold sores on lips	$\frac{D}{2} \frac{100}{7} \frac{S}{S} \frac{V}{0}$
	Phagoviridae (naked)	T-phages		$\frac{D}{2} \frac{30}{50} \frac{X}{X} \frac{B}{0}$
Ribohelica (helical)	Rigidoviridae (naked)	Tobacco mosaic virus	Tobacco mosaic disease	$\frac{R}{1} \frac{2}{5} \frac{E}{E} \frac{P}{0}$
	Orthomyxoviridae (enveloped)		Influenza	$\frac{R}{1} \frac{\Sigma 5}{1} \frac{S}{E} \frac{V}{0}$

Appendix Virus classification (*cont.*)

Phylum: *Vira*
Subphylum: *Deoxyvira*

Class	Family	Genus	Diseases and common viruses	Family cryptogram
Ribohelica (helical)	Paramyxoviridae (enveloped)	Paramyxovirus	Mumps, Newcastle disease, Sendai, Measles, Canine distemper, Rinderpest	$\dfrac{R}{1}\,\dfrac{7}{1}\,\dfrac{S}{E}\,\dfrac{V}{0}$
	Rhabdoviridae (enveloped)	Rhabdovirus	Vesicular disease virus, Rabies	$\dfrac{R}{1}\,\dfrac{4}{2}\,\dfrac{U}{E}\,\dfrac{V,I}{O,Di}$
	Retroviridae	Oncovirus	Rous sarcoma virus	$\dfrac{R}{1}\,\dfrac{\Sigma 10}{2}\,\dfrac{S}{E}\,\dfrac{V}{0}$
Ribocubica (cubical)	Napoviridae (naked)	Napovirus	Turnip yellow mosaic virus (TYMV)	$\dfrac{R}{1}\,\dfrac{2}{30}\,\dfrac{S}{S}\,\dfrac{P}{0}$
	Picornaviridae (naked)	Enterovirus, Rhinovirus, Calicivirus, Aphthovirus, Cardiovirus, Equine rhinovirus	Polio, Common cold, Foot-and-mouth, EMC	$\dfrac{R}{1}\,\dfrac{2.6}{30}\,\dfrac{S}{S}\,\dfrac{V}{0}$
	Androphagoviridae (naked)	Androphagovirus	MS2, R17 bacteriophages	$\dfrac{R}{1}\,\dfrac{1.5}{30}\,\dfrac{S}{S}\,\dfrac{B}{0}$
	Reoviridae (naked)	Reovirus		$\dfrac{R}{2}\,\dfrac{\Sigma 15}{15}\,\dfrac{S}{S}\,\dfrac{V,P}{0}$
	Togaviridae (enveloped)	Alphavirus	Sindbis	$\dfrac{R}{1}\,\dfrac{4}{7}\,\dfrac{S}{S}\,\dfrac{V,I}{Di}$
		Flavivirus	Yellow fever	$\dfrac{R}{1}\,\dfrac{4}{7}\,\dfrac{S}{S}\,\dfrac{V,I}{Di}$

Further reading

The following books are suitable texts for background reading on relevant biochemical topics and for further study in virology:

Adams, R. L. P., Burdon, R. H., Campbell, A. M. & Smellie, R. M. S. (revisers) (1976). *Davidson's The Biochemistry of the Nucleic Acids*, 8th edn. Chapman & Hall, London.
A comprehensive introduction to nucleic acids.
Dickerson, R. E. & Geis, I. (1969). *The Structure and Action of Proteins.* Harper & Row, New York, Evanston & London.
An excellent introduction to protein chemistry.
Harrison, R. & Lunt, G. G. (1975). *Biological Membranes. Their Structure and Function.* Blackie, Glasgow & London.
Watson, J. D. (1976). *Molecular Biology of the Gene*, 3rd edn. W. A. Benjamin, Inc. New York.
A well-illustrated and up-to-date survey of molecular biology.
Haggis, G. H. (1968). *The Electron Microscope in Molecular Biology.* Longmans, Green & Co., London.
A useful introduction to the variety of techniques used in electron microscopy of viruses and cells.
Roitt, I. (1975). *Essential Immunology*, 2nd edn. Blackwell, Oxford.
A well-illustrated introduction to immunology.
Fenner, F., McAusland, B. R., Mimms, C. A., Sambrook, J. & White, D. O. (1974). *The Biology of Animal Viruses*, 2nd edn. Academic Press, New York & London.
A comprehensive standard text. A 'must' for anyone intending to study virology more deeply.
Burke, D. C. & Russell, W. C. (Ed.) (1975). *Control Processes in Virus Multiplication.* Symposium 25 of the Society for General Microbiology. Cambridge University Press, London.
Suitable reading for more advanced students.
Fraenkel-Conrat, H. & Wagner, R. R. (1974 onwards). *Comprehensive Virology.* Plenum Press, New York & London.

A new series of up-to-date monographs on all aspects of virology; about 22 volumes are planned.

Dalton, A. J. & Haguenau, F. (Ed.) (1973). *Ultrastructure of Animal Viruses and Bacteriophages. An Atlas.* Academic Press, New York & London. Well illustrated with excellent electron micrographs.

Index

actinomycin D, 112, 135
adenovirus
  antigens, 68
  proteins, 64, 68
  symmetry, 57
affinity chromatography, 26
agar gels, 27
agglutination, 29
amantadine, 132
animal viruses
  bovine enterovirus, 31, 59, 61
  encephalomyocarditis (EMC) virus, 113
  foot-and-mouth disease virus (FMDV), 13, 19, 35, 83, 117
  herpes virus, 92
  influenza virus, 11, 13, 30, 72, 135
  measles virus, 13, 26, 69, 84
  molluscum contagiosum, 121
  mumps virus, 13
  Newcastle disease virus (NDV), 13, 69, 135
  papilloma virus, 11, 121
  polyoma virus, 92, 121
  rabies virus, 13
  reovirus, 80, 101
  rhinovirus, 19
  scrapie virus, 10. 18
  vaccinia virus (smallpox), 1, 14, 72, 103, 134, 135
  vesicular stomatitis virus (VSV), 14, 72, 99
antibodies, 19, 24, 25, 26, 27
antigens, 19, 21, 24, 25, 26, 27
attachment
  of animal viruses, 78
  of bacteriophages, 76, 77
autoimmune, 128, 131
autointerference, 126
autoradiography, 26

bacteriophages
  acquisition of new metabolic activity by cells infected by, 90
  attachment, 76
  composition, 5
  discovery, 4
  DNA bacteriophages: T-even phages, 5, 73, 77, 85, 88, 90, 105; T-odd phages, 73, 77; ϕX 174, 67, 75, 78, 106; lambda (λ) phage, 106, 108, 109, 110, 111
  inhibition of cell metabolism by, 88
  integration, 110
  intemperate phages, 108
  in vitro replication, 9, 116
  lysogenic, 109
  maturation of T-phages, 124
  messenger RNA, 85
  number of genes, 92
  penetration, 78
  prophage state, 83
  reconstitution, 53
  RNA bacteriophages: R17, 92, 94, 95, 112, 114; f2, 94; Qβ, 94, 115
  sigma factors, 92
  size, 34
  structure, 73
  titration, 21
  transduction by, 111
  yield, 8
biological amplification, 21
biosphere, 3, 76, 136
bovine enteroviruses, 31, 59
budding processes, 39, 69

cancer, 9, 122
capsid, 12, 40, 43, 44, 53, 54
capsomers, 11, 52, 65
cells
  fusion, 71
  lysis, 10
  transformation, 10, 121, 122, 136

141